Everyday Peace?

RGS-IBG Book Series

For further information about the series and a full list of published and forthcoming titles please visit www.rgsbookseries.com

Published

Everyday Peace?

Politics, Citizenship and Muslim Lives in India

Philippa Williams

WILEY Blackwell

This edition first published 2015
© 2015 John Wiley & Sons, Ltd.

Registered Office
John Wiley & Sons, Ltd, The Atrium, Southern Gate, Chichester, West Sussex, PO19 8SQ, UK

Editorial Offices
350 Main Street, Malden, MA 02148-5020, USA
9600 Garsington Road, Oxford, OX4 2DQ, UK
The Atrium, Southern Gate, Chichester, West Sussex, PO19 8SQ, UK

For details of our global editorial offices, for customer services, and for information about how to apply for permission to reuse the copyright material in this book please see our website at www.wiley.com/wiley-blackwell.

The right of Philippa Williams to be identified as the author of this work has been asserted in accordance with the UK Copyright, Designs and Patents Act 1988.

Library of Congress Cataloging-in-Publication Data

Williams, Philippa (Lecturer), author.
  Everyday peace? : politics, citizenship and Muslim lives in India / Philippa Williams.
    pages   cm – (RGS-IBG book series)
  Includes bibliographical references and index.
  ISBN 978-1-118-83781-8 (cloth) – ISBN 978-1-118-83780-1 (pbk.)   1. Peace.
2. Muslims–India–Uttar Pradesh.   3. Hindus–India–Uttar Pradesh.   I. Title.
II. Series: RGS-IBG book series.
  JZ5584.I4W55 2015
  303.6'609542–dc23
                                                                2015017218
A catalogue record for this book is available from the British Library.

Cover image: © Philippa Williams

Set in 10/12.5pt Plantin by SPi Global, Pondicherry, India

The information, practices and views in this book are those of the author(s) and do not necessarily reflect the opinion of the Royal Geographical Society (with IBG).

Printed and bound in Malaysia by Vivar Printing Sdn Bhd

1   2015

For my parents, Maureen and Ted

Contents

Series Editors' Preface

The RGS-IBG Book Series only publishes work of the highest international standing. Its emphasis is on distinctive new developments in human and physical geography, although it is also open to contributions from cognate disciplines whose interests overlap with those of geographers. The Series places strong emphasis on theoretically-informed and empirically-strong texts. Reflecting the vibrant and diverse theoretical and empirical agendas that characterize the contemporary discipline, contributions are expected to inform, challenge and stimulate the reader. Overall, the RGS-IBG Book Series seeks to promote scholarly publications that leave an intellectual mark and change the way readers think about particular issues, methods or theories.

For details on how to submit a proposal please visit:
www.rgsbookseries.com

David Featherstone
University of Glasgow, UK

Tim Allott
University of Manchester, UK

RGS-IBG Book Series Editors

Acknowledgements

The ideas for this project have changed a great deal over the years, for which I am extremely indebted to a large number of people for their generosity, insight and intellect. The vast majority of research on which this book is based was conducted between 2006 and 2008 during my PhD studies at the School of Geography, and further developed on fieldtrips in 2010 and 2011 when I was a Smuts Research Fellow in Commonwealth Studies at the Centre of South Asian Studies, both at the University of Cambridge.

First and foremost I am hugely grateful to the residents of Varanasi and in particular, Madanpura, for generously accepting me as a fairly constant presence in their lives for over 14 months. Abdul Ansari, Mohammed Toha, Atiq Ansari and Rana P. B. Singh were just some of those who facilitated my research in different ways, whilst it was a conversation with Manju Vira Gupta that led me to Varanasi in the first place. My utmost thanks to my Research Assistant, Ajay Pandey, whose friendship, curiosity and dedication made the research process all the more productive and enjoyable. I also acknowledge the research assistance of Hemant Sarna, Afreen Khan, Pintu Tripathi and I. B. Misha at different stages in the project.

In a notoriously overwhelming and bustling Indian city I found a wonderful sanctuary in the home away from home of Prabhudatt, Pintu, Shivangi and Babu Tripathi, from whom I learned so much about Varanasi and life more generally. My thanks to Simon Roberts for such a precious introduction, and to other companions and occasional informants in the city: Vinay, Rakesh, Navneet and Nitya as well as Jolie Wood, Mike Thompson and Brita Ahlenius. My Hindi teacher, Virendra Singh, was also hugely instrumental in shaping my introduction to, and freedom in, the city.

In Cambridge, I wish to acknowledge the long-standing support and instrumental nudge by Phil Howell which kick-started my graduate career. My warmest thanks to my PhD supervisor Bhaskar Vira, whose constructive comments and encouragement have been hugely important, in ways that go far beyond this book. A 3-month ESRC overseas institutional visit to the University of Seattle, Washington proved to be formative in furthering my academic career and continuing interest in South Asia. It was both in the USA and subsequently on my return to the UK that I have benefited enormously from conversations with Craig Jeffery, Jane Dyson, Dena Aufseeser and Stephen Young, whose comments greatly improved the final draft of the manuscript.

I am grateful for the constructive comments of my PhD examiners, Roger Jeffery and Sarah Radcliffe. As a PhD student and Research Fellow I was lucky enough to meet superb friends, colleagues and collaborators who formed a stimulating research environment. There are too many to mention, but ideas for this book have undoubtedly been shaped by discussions with Deepta Chopra, Heather Bedi Plumridge, Tatiana Thieme, Fiona McConnell, David Beckingham, Eleanor Newbigin, Delwar Hussain, Karenjit Clare, Jesse Hohman, Venkat Ramanujam Ramani, Emma Mawdsley, Gerry Kearns and Humeira Iqtidar.

It was during my time at the Centre of South Asian Studies that I started reworking aspects of the PhD into a book. I benefited enormously from the institutional and usually caffeinated support of Chris Bayly, Barbara Roe, Kevin Greenbank and Rachel Rowe.

In recent years I have worked especially closely with Nick Megoran and Fiona McConnell as well as contributors on the production of an edited volume on the Geographies of Peace for I.B. Tauris. Talking and writing together about peace has undoubtedly shaped and sharpened my conceptual approach to peace, and challenged me to think about the geography of peace in other places.

Sections of this manuscript have been shared at various conferences and seminars, in Oxford, Madison, Seattle, Cambridge, Washington, Yale, Leicester, Manchester, Boston and Princeton, where it has benefited from the comments of amongst others, Alpa Shah, Ed Simpson, Sara Koopman, Manali Desai, Nosheen Ali, Glyn Williams, Isabel Clarke-Deces, Tim Raeymaekers, Gurharpal Singh and Barbara Harriss-White.

The editing for this book has taken place since starting a new lectureship at Queen Mary University of London. I wish to thank the third-year students taking my module, 'Contemporary India: Society, politics and the economy' for energizing the editing process, and to acknowledge the support of colleagues, Al James, Miles Ogborn, Cathy McIlwaine and Kavita Datta; I look forward to new academic ventures at QMUL.

The manuscript has benefited tremendously from the critical comments on the book proposal and manuscript of the anonymous reviewers and the editor, Neil Coe, for which I am very grateful. Neil Coe has been an excellent editor with swift, insightful and clear comments on the editing process. Many thanks, too, to Jacqueline Scott for cheerfully and efficiently keeping things on track.

I would like to acknowledge the generous financial support I have received over the past decade from a number of bodies including the Economic and Social Research Council, Smuts Commonwealth Fund and British Academy small grant, without which fieldwork in India, as well as participation at conferences, would not have been possible.

The biggest thank you is saved for my family, in particular, my parents Maureen and Ted, and sister Catherine, whose unconditional support, sound advice and unreserved affection has made this book possible. And to my husband, James for believing in this project from the beginning, tolerating my long stints away and providing encouragement and advice when it was most needed. And finally, I am grateful to the arrival of our son, Alexander, who has brought additional happiness, love and perspective to academic writing.

Philippa Williams

London, 2014

An earlier version of Chapter 3 appeared as Williams, P. (2007) "Hindu Muslim Brotherhood: Exploring the Dynamics of Communal Relations in Varanasi, North India." *Journal of South Asian Development* 2(2): 153–76.

A version of Chapter Four appeared as Williams, P. (2013) "Reproducing Everyday Peace in North India: Process, Politics and Power." *Annals of the Association of American Geographers* 103(1): 230–50.

Chapter Six draws on aspects of two articles: Williams, P. (2012) "India's Muslims, Lived Secularism and Practicing Citizenship." *Citizenship Studies* 16(8): 979–95 and Williams, P. (2011) "An Absent Presence: Experiences of the 'Welfare State' in an Indian Muslim *mohallā.*" *Contemporary South Asia* 19(3): 263–80.

List of Abbreviations

BHU	Banaras Hindu University
BIMARU	Bihar, Madhya Pradesh, Rajasthan and Uttar Pradesh
BJP	Bharatiya Janata Party
BSP	Bahujan Samaj Party
BJP	Bharatiya Jan Sangh
CM	City Magistrate
DM	District Magistrate
DPSP	Directive Principles of State Policy
DSP	Deputy Superintendent of Police
GSC	Golden Sporting Club
FDI	Foreign Direct Investment
HIS	Health Insurance Scheme
INC	Indian National Congress Party
INTACH	Indian National Trust for Art and Culture
NGO	Non governmental organization
MLA	Member of Legislative Assembly
MP	Member of Parliament
NDA	National Democratic Alliance
OBC	Other Backward Class
PAC	Provincial Armed Constabulary
POTA	Prevention of Terrorism Act

PUCL	People's Union for Civil Liberties
PUHR	People's Union of Human Rights
RAP	Rapid Action Force
RSS	Rashtriya Swayamsevak Sangh
SC	Scheduled Classes
SEZ	Special Economic Zone
SGDP	State Gross Domestic Product
SP	Samajwadi Party
SSP	Special Superintendent of Police
ST	Scheduled Tribes
UC	Upper Caste
UP	Uttar Pradesh
UPA	United Progressive Alliance
VDA	Varanasi Development Authority
VHP	Vishva Hindu Parishad

Glossary

Ansāri	Muslim caste/community, often synonymous with the occupation of weaving, self-referential term for Muslim weavers.
Ante	Large spinning structure for ordering silk yarn in preparation for the loom
antewāllā	Sari production worker who spins silk yarn in preparation for winding *tānā* on to *belans*
balvā ārti	A Hindu worship ceremony in which lighted lamps are moved in a circle before the image of the deity
bānā	Silk weft, runs across the width of the handloom
bakha khani	Oven-baked sweet prepared with dates, eaten during ifta
Bhagavad Gītā	Revered Sanskrit Hindu scripture
Bakra-Eid	Muslim festival
belan	Lit: roller. A large wooden bail on to which the *tānā* is rolled and attached to the loom ready for weaving
balvā	Riot, disturbance
Bhakti	A devotional Hindu movement which promoted caste equality and Hindu-Muslim unity
bhāī	Brother
bhāī-bhāī	Brotherhood
bhaīachārā	Brotherhood

bīmār	Ill, unwell
birādarī	Kinship; community
burqā	A loose, usually black or light blue robe worn by Muslim women
būttī kārwa	Intricate pattern work done on the handloom, often by children
chāy	Tea
chhajjā	Awning, for protection from rain or sun
chunnī lehengā	Traditional long decorative skirt and stole worn with corset
crore	Ten million
dāg	Mark
dalāl	Commission agent/tout
Dalit	Lit. 'crushed and broken', term for people traditionally regarded as untouchables. Although the caste system has been officially abolished *dalits* still experience societal prejudice. As an administrative category *dalits* are given rights to employment and education in state institutions
dangā	Riot
dārhī	Beard
dhāgā	Thread
dhobhī	Washerperson; caste group whose traditional occupation was to wash other people's clothes
dum	Pride
dupattā	Thin, long shawl that accompanies the a *shalvār kamēz*
Gangā Jamuni tahzēb/sanskritī	Lit. Ganges–Jamuna culture. A euphemism for the mutually participatory coexistence of Hindu and Muslim culture in north India
galī/s	Narrow lane/s
Gaddī	Shop or showroom
gaddidār	Sari trader, large-scale transactions
ghāt	Riverbank/steps down to the river
girhasta	Sari trader, smaller scale than *gaddidar*, often with personal weaving experience
gosāīns	Mendicant trader-soldiers
harām	Forbidden
havelī	Lit. 'enclosed place', private mansion/house
Hindutva	Lit. 'Hinduness', political and cultural project that advocates Hindu nationalism
īfta	Meal taken at sunset to breakfast during the month of *Ramzān*

Imām	A minister or priest who performs the regular service of the mosque
īmān	Truth, honesty
imāmbārā	Large tomb, buildings or rooms used for Islamic ceremonies
isha'a	One of the five daily Muslim prayers, takes place at night
julāhā	Term given to weavers, corresponds with one of the lowest Hindu castes, regarded by weavers as a derogatory identity
kārkhānā	Workshop
kārobāri	Businessmen
khoyā	Milk product, used as a base for Indian sweets amongst other foods
kurtā-pyjāmā	Knee-length shirt worn over straight trousers
kutchā	Poorly constructed
lākh	Hundred thousand
lāthī	Wooden sticks, often used as weapons by the police
lapetnawāllā	Sari production worker, rolls *tānā*/warp on to *belans* ready for the handloom
madrasā	Islamic school
Mahājan	Sari trader, businessman
Mahant	Chief priest of a *mandir*
Mahābhārata	Sanskrit epic from ancient India, forms an important part of Hindu mythology
maidān	Large open space
majmā	Crowd
mallā	Boatman; caste group whose traditional occupation was to transport goods and people by boat
Māmā jī	Maternal uncle
mandī	Market
mandir	Temple, Hindu place of worship
masjid	Mosque, Muslim place of worship
Maulānā	Islamic teacher; general term of respect for
melā	Festival
mohabbat	Love, affection
mohallā	Neighborhood
moksha	Rebirth
Mufti	Premier Muslim scholar who interprets the *Sharia*
mūrti	Statue
nagādā	Drum
namāz	Muslim prayers

namkīn	Salted snacks
pān	Betel leaves
pakkā	Well made, proper
pandāl	Huge tent erected to house religious festivals
pardā	Lit. 'curtain', refers to the predominantly Muslim custom of keeping secluded from the public
pattīwāllā	Worker who punches the holes in jacquard card through which the silk pattern is transmitted on to the handloom
phandā	Noose
prasād	Sweets, as offering to gods
pūjā	Ritual of worship
rajnitī	Politics
Ramāyana	Sanskrit epic from ancient India, forms an important part of Hindu mythology
Ramzān	Muslim month of fasting
reshmi	Silk
rozā	Fast
rozi-rotī	Daily bread; livelihood
rāshtra	Nation
saptapuri	Seven cities of Hindu pilgrimage
Sādhū	Hindu ascetic
shakti	Power
Shariā	Muslim code of law derived from the Qur'an
shalvār kamēz	Loose fitting trousers worn with long shirt
Scheduled Class (SC)	An administrative group of people traditionally thought of as untouchable. Given right to preferential treatment in employment and education in state institutions, as well as in electoral politics
Scheduled Tribe (ST)	An administrative group of people. Also called 'Adivasi'. Given right to preferential treatment in employment and education in state institutions, as well as in electoral politics.
sparklewāllā	Worker who applies glitter on to saris
suti	Cotton
tamāshā	Spectacle
tānā	Silk warp, runs the length of the handloom
tānejorna	Worker who attaches the old silk *tānā*/warp to the new silk *tānā*/warp
tāziyā	Representational tombs of the Karbala martyrs, processed during Muharram

tilak	Orange or red mark applied to the forehead following a *pujā* or in an engagement ceremony
topī	Small cap, often refers to the cap worn by Muslim men
ulās	Courage
vyāparī	Businessman
waqf	Charitable endowment, gift of land or property made by a Muslim; intended for religious, educational or charitable use
yātrā	Journey
zarī	Thin threads of gold or silver used in weaving and embroidery

List of Figures

Chapter One
Introduction

Situating Everyday Peace

As we sat on the padded cotton-lined floor of Jamaal's sari showroom, drinking Sprite out of thick green glass bottles, he proceeded to tell me with great enthusiasm just how intertwined relations were between Muslims and Hindus in the silk sari industry of his north Indian city. "Hindus and Muslims all work in this industry," he said "it's like a cycle, the silk is sold by a Hindu and prepared by a Muslim, then the weaver is Muslim and the embroidery is later done by a Hindu and finally the sari is sold to a Hindu woman who wears it." Jamaal was a member of a prosperous Muslim family, but as many Muslim and Hindu informants had told me before our meeting, and as I would continue to hear many times after this encounter: "in the silk sari culture *itna pakka rishtedar hai*" (relations are really good), they are like "*tānā bānā*" (warp and weft). The reiteration of this narrative, often framed in terms of "brotherhood" and consciously describing everyday harmonious relations between Hindus and Muslims, is striking in the context of north India.

A glance at the Indian and western print-media over the past few decades portrays a different story; of the incommensurable differences between

Everyday Peace?: Politics, Citizenship and Muslim Lives in India, First Edition. Philippa Williams.
© 2015 John Wiley & Sons, Ltd. Published 2015 by John Wiley & Sons, Ltd.

Hindus and Muslims, and the persistently violent nature of their interactions, often made manifest in "riots" (see Tambiah 1996). But Jamaal was not an idealist. He acknowledged that there were sometimes tensions between these two communities, and in recent years riots had created challenges for their working relationships, which had subsequently prompted different parties to actively ensure that the "*shanti*" (peace) was not disturbed. Such "peace talk" was not only articulated by Muslim weavers and businessmen eager to make a living in a challenging market, but it also animated conversations amongst other residents about the character of their city, and it underpinned the political and media discourses that were deployed about urban life, especially where everyday peace appeared to be under threat.

Sensitive to the rhetorical and lived potency of "peace talk" and practice in this north Indian city, this book develops a critical and grounded approach to understanding everyday peace. The account presented here is concerned with questions about what peace looks like, how peace is spatially and socially produced and reproduced, in whose image is peace constructed and how different people experience peace differently. Importantly, it seeks to interpret experiences of peace from the margins, and accordingly develops an understanding of everyday peace through the lives of those living in a Muslim neighborhood, in the majority-Hindu city of Varanasi.

India's Muslims, as in other parts of the world, are regularly marked either as "dangerous terrorists" or as "passive victims" and in both instances subjected to patterns of discrimination and stigma. A key argument of this book is that these pervasive narratives differentially conceal the notion of the Indian Muslim as citizen. They obscure the ongoing struggles by Muslims to improve or maintain their material wellbeing as well as the myriad ways in which Muslims are orientated towards securing and maintaining peace within the Indian secular state and social milieu. By grounding peace in this specific spatial and social context, this book illustrates how peace interacts with agency and legitimacy, citizenship and justice, and how these are constitutive of the production of place. As such, this is a book about the geography of peace: how peace makes place, and how a place makes peace. It demonstrates how as an inherently relational construction, peace is both the product of and the context through which differences and connections are assembled and negotiated across scale, articulated through different forms of "peace talk" and informed by uneven geographies of power. Rooted in the local, this particular geography of peace is thus concerned with how peace is socially and spatially (re)produced in and through interconnected sites and scales, including the body, the neighborhood, the city, region and nation.

India represents a fascinating place within which to problematize the notion and practice of peace because of the region's postcolonial experience of religious politics, violence and nonviolence. Incidents of Hindu–Muslim violence, notably referred to as "riots," have taken place in India, during and since the subcontinent's partition, and increased in their frequency and intensity between the late 1980s and 2000s (e.g. Pandey 1990; Roy 1994; Nandy 1995; Kakar 1996; Tambiah 1996; Oza 2006). More recently, Islamist terrorist attacks in India's metropolitan and regional cities have caused death and injury (Ahmad 2009; Bishop and Kay 2009). The notion of immanent tension and violence between Hindus and Muslims is also reflected at the geopolitical scale with respect to the hostile nature of India–Pakistan relations, for instance, political posturing around nuclear capabilities, the ongoing dispute over Kashmir and following the 2008 Mumbai terrorist attacks (see Wirsing 1998; Ganguly and Hagerty 2005; Ramesh et al. 2008). Meanwhile, the 2000-km fencing project along India's national border with Bangladesh represents a particularly tangible geopolitical articulation of the divisive nature of Hindu–Muslim politics that informed the making of India and continues to shape politics in the region (see Jones 2009; Hussain 2013). These tensions have become intertwined in the making and remaking of India's geopolitical borders and politics and ultimately, underpin its citizenship project.

It is not just within India's cities and along its borders that the nation's religious differences and inherited inequalities find expression, but also in state policy spaces, through the drafting and interpretation of the Indian Constitution. The Constitution was underpinned by the need to imagine how the idea of India's democracy would be institutionalized and practiced in order for India to realize a peaceful postcolonial future. Key challenges facing the architects of the Constitution were: how to manage and minimize imagined and material differences between India's majority Hindu population and its largest religious minority, Muslims; and how to create mechanisms for the inclusion of marginalized groups in order to avert the possibility of separatism or dissent and further inevitable violence. Given this, I argue that the Indian state policy of secularism can be interpreted as a living document for national peace, one that is informed by the understanding that to reproduce peace, spaces of tolerance, freedom, respect and equality need to exist between different communities. And, the state has to remain equidistant from all religions in its style of politics, although not entirely absent as is the case in most European applications of secularism. Despite the fact that secular practices of respect and equality between all religions have been compromised, with sometimes violent outcomes between religious groups, the idea of secularism nonetheless persists as a powerful rhetorical imaginary mobilized across different scales. Today, secularism constitutes a dominant political narrative and practice at different scales within Indian society. Exploring how it constitutes a vehicle for hope and opportunity for marginalized religious groups to make their claims to the nation and their rights to equality and justice importantly contributes to understandings of "lived secularism" as citizenship from below, and lived practices of peace.

When it comes to exploring Indian Muslim experience in India today, especially in the context of relational life, it is difficult not to become immersed in dominant literatures that emphasize the role of religious division and violence in shaping India's national and geopolitical relations. Without denying the import of critiques concerning moments of intercommunity and international strife, the prevalence of such views has unwittingly functioned to construct and to fix Hindu–Muslim interactions as intractably antagonistic and habitually violent. More importantly, the focus on violent events means that actual lived realities in much of India, characterized by inter-community coexistence and everyday peace, risk being occluded.

Interestingly, this principal focus on violence and divisive politics is not unique to India. More generally, matters of war and violence have proven to be far more seductive foci for academic analysis whilst issues of peace have been typically relegated to a position of referential obscurity, in which "peace" is constructed simply as a negative, empty state, as the absence of violence. On the idea of nonviolence Kurlansky (2006) suggests that the problem is that despite the concept being the subject of praise by every major religion, and practitioners of nonviolence recognized throughout history, *violence* is taken for granted as the fundamental human condition (Arendt 1969). The implicit yet seldom expressed viewpoint amongst most cultures is that violence is real and nonviolence is unreal (Kurlansky 2006, p. 6).

This book argues the case for peace and nonviolence as something "real" and worth understanding, in its own right. Whilst we revere war we have not been taught how to think about peace (Forcey 1989). This may partly explain why previous engagements with peace in geography (e.g. Stolberg 1965; Pepper and Jenkins 1983; Kliot and Waterman 1991; Flint 2005b) and anthropology (e.g. Howell and Willis 1989; Sponsel and Gregor 1994) have not produced a coherent and sustained body of scholarship, let alone a critical perspective on peace. By contrast, the social sciences have contributed to deconstructing and uncovering the nature of power relations inherent in the course of "war," particularly the War on Terror (Gregory 2006; Pain and Smith 2008; Dodds and Ingram 2009), conflicts in the Middle East (Yiftachel 2006) and the Balkans (Dahlman 2004). Even where "peace" is the focus of study in geography (Kobayashi 2009; Gregory 2010) and in peace studies (Barash 2000; Cortright 2008), the concept and actual experience of peace itself has been given remarkably little attention. However, the arguments for a more critical examination of peace have been building within different disciplines over recent years (Richmond 2005, 2008; Jutila et al. 2008; Williams and McConnell 2010; Megoran 2010, 2011 and McConnell et al. 2014).

Responding to these arguments this book evokes the pragmatism demonstrated by Jamaal as he seeks to negotiate peaceful relations in Varanasi's silk market. Peace is not regarded as a trouble free product, but instead an ongoing process that is at once political and infused with power across different sites and scales. Adopting a geographical approach to notions of space, place and politics this book makes four key theoretical interventions. First, by grounding peace in place, this account illuminates the role of human agency in the (re)production of everyday peace. My account shows how local actors actively negotiate and (re)produce peace as policy, narrative, practice and strategy within different urban spaces and across different scales. Paying attention to local agency echoes recent moves by Oliver Richmond to foreground the role that international *and* local actors can play in peace keeping, through a hybrid local-liberal peace (Richmond 2010). In thinking about everyday peace I expand the focus beyond postconflict spaces, and into more prosaic contexts to show the contingent ways in which individuals and groups are situated within wider arenas that inform the everyday possibility and practice of living together.

This focus also develops scholarship on India from political science perspectives that have examined the role of agency in escalating intercommunity violence (Brass 2003) and wider societal mechanisms underpinning peace (Varshney 2002), but which have not directly addressed questions of peace and agency and the micropolitics of everyday life. Understanding the micromechanisms that constitute peace helps to explain how and why actors differentially orientate themselves towards people of difference through contrasting experiences of tolerance, solidarity, hospitality, indifference, tension and brotherhood, and how these everyday realities impact the shifting potential to (re)produce peace "on the ground." With agency also comes questions of responsibility and legitimacy, so not only why and how certain people chose to act towards peace but also whether their actions are positively recognized, or not, and by whom.

Secondly, I contend that peace is political. Whilst peace is more often portrayed as a condition that is without or after politics and violence, the empirical narrative that emerges here is that peace is, in and of itself, political. A critical approach to peace is paramount for unpacking in whose image peace is made and remade, who bears the responsibility for maintaining peace and what the implications of this are for future peace. Examining the interacting role of narrative and practice in the (re)production of peace reveals the political work of peaceful narratives, for, whilst they may appear universally "good" they may also conceal and perpetuate uneven relations of power and marginality. Critical scholars of international relations and geographers, too, have exposed the structures of power that underpin the discourse of "liberal peace," with its particular "Western" vision for what peace should look like and how it should be done (Duffield 2002; Jeffrey 2007; Richmond

2008; Springer 2011; Daley 2014). The book shows how other expressions of "peace talk" such as "brotherhood," intercommunity harmony and interdependence, along with its vernacular equivalents, act as powerful narratives in both describing and constructing peaceful realities across different scales. But also, that these narratives have particular histories and are constitutive of a particular politics. Placing peace, therefore, foregrounds how power is integral to the making, shaping, undoing and remaking of peace. Where power is recognized as the immanent normalizing force that operates through the detailed fabrics of people's lives, it is important to consider the different techniques of peace and how these may include domination, coercion, seduction, authority, compliance and complicity and not just, or not always, involve the reconciliation of tensions and realization of equality and justice that is often imagined as integral to "peace." Whilst inspired by Johan Galtung's rich conceptualization of peace, the book therefore challenges the binary between "negative peace" as the absence of violence and "positive peace" as the presence of social justice (Galtung 1969).

Third, the book develops an understanding of peace as process, as always becoming. From this perspective, peace demands ongoing labor and work rather than standing as an endpoint or as something which can be concluded (see Cortright 2008). As such, peace is precarious, it is contingent and its contours are continuously being reworked by actors and events along a continuum from the local to the global. Bearing in mind the political work involved in peace as process, the book questions assumptions that peace is a generative and positively transformative process (see Galtung 1969) and argues that practices of peace can also re-inscribe patterns of marginality. This approach builds on Laura Ring's (2006) problematization of peace in a Karachi apartment building, and intersects with an understanding of human agency which goes beyond resistance, to consider the ways in which pragmatism, acceptance, resilience and patience find expression in efforts to make and re-make everyday peace. As such, peaceful realities need to be interpreted within a framework that encompasses broader practices of coexistence, as well as practices of citizenship, where subaltern actors struggle for inclusion in the city and wider polity. Finally, this account departs from "top down" conceptualizations of peace to represent a situated geography of peace. This in part responds to recent calls within peace studies to interpret the meaning of "peace" through fieldwork (Jutila et al. 2008) and extends anthropology scholarship on "peaceful sociality" (Howell and Wills 1989; Sponsel and Gregor 1994) by thinking more explicitly about the spatial and scalar processes that constitute peace. The book therefore attends to everyday peace "on the ground" to show how practices and narratives of peace are spatially situated and socially (re)produced within a particular north Indian city. It shows how peace as "talk" and as practice informs the making of place, and that, in turn place makes for particular forms of peace. It demonstrates the multi-scalar practices of different actors and demonstrates the interconnections between everyday peace and related practices of citizenship and experiences of justice from a subaltern perspective.

In order to develop this grounded approach to "everyday peace" the book centrally draws on different scalar narratives and practices of peace in and around a Muslim neighborhood in Varanasi, a regional city in north India and home to one million people. On the one hand Varanasi represents an "ordinary city" which contrasts with the metropolitan centers of New Delhi and Mumbai that are more often in the academic limelight, owing to their discernible global reputations and interconnections (see Robinson 2002). But, on the other hand Varanasi is also an "extraordinary" city; as an important center for silk manufacture involving Hindus and Muslims, who comprise almost one-third of the population, as a sacred Hindu pilgrimage site, and also as a city locally renown to be a "microcosm of India" given its diverse representation of other people from all over India. In many ways because of its "extraordinary" character, Varanasi is popularly constructed in regional and national imaginations as a "peaceful" city. Understanding how "peace talk" is expressive of various scalar articulations of Hindu-Muslim interactions and also contributes to a (re)making of place as peaceful is a central objective of the following chapters.

In order to conceptualize and empirically illuminate an understanding of everyday peace in north India the remainder of this chapter brings into dialogue diverse but related collections of literature over six core sections. First, I briefly introduce the politics of Hindu–Muslim violence and nonviolence in India to contextualize the theoretical and empirical impetus for research on everyday peace. Secondly, I outline the contours of studies on peace and peaceful sociality from different disciplinary spheres before turning more explicitly to look at geographies of peace in the third section. Fourth, I show how feminist geopolitics provides a useful framework through which to interpret everyday peace in a way that foregrounds the role of subaltern agency, space and the political. In addition, I explore how scholars have approached questions of violence and the everyday. In the fifth section I extend the scalar politics of feminist geopolitics to illustrate how citizenship affords a complementary way to theorize agency and practice from within the margins in the context of state and society relations. The sixth section turns to work on Muslim geographies and introduces the situation of India's Muslims more specifically, by drawing on literature in South Asian Studies. The chapter concludes with an overview of the book's structure and its key arguments.

The Politics of Hindu–Muslim Violence and Nonviolence

The partition of the Indian subcontinent in 1947 led to the creation of the secular, multicultural nation of India, with a sizeable majority Hindu population, and the "Muslim state" of Pakistan. The act of partition was realized through months of intense material and rhetorical violence that took place across the Indian subcontinent, as India's Muslims and Pakistan's Hindus moved to be on the "right" side of the border, and through this process the perception of religious difference became reified (see Pandey 2001; Khan 2007). Ironically, more Muslims remained in India than settled in Pakistan, but as the largest religious minority they have experienced practices of discrimination in economic, educational and political life as well as physical violence since Indian independence (Hasan 2001). Such realities have undermined their potential to fully realize their formal citizenship rights, as I discuss later. The position of India's Muslims has been heavily influenced by the transformation of Hindu nationalism in contemporary India (Jayal 2011). The central objective of Hindu nationalism is to create political unity amongst the Hindus, which requires the imaginary of an ethnically homogeneous community with a singular Indian citizenship (Jaffrelot 1996). Thomas Blom Hansen has documented the contours of this cultural–political movement to show how "the identification of the Other as Muslims is instituted and repeated endlessly by Hindu nationalism" (1996, p. 152). Similarly, Arjun Appadurai (2006) argues that the exorcizing of the Muslim Other is symptomatic of the majority's "anxiety of incompleteness" – that is, the feeling that the minority, however small, is hindering the realization of a pure and untainted national ethos.

With the rise of popular Hindu nationalism during the mid-1980s to early 2000s, Hindu–Muslim "riots" have taken place with increased intensity and frequency, more often in parts of urban north India. Rupa Oza (2009) has documented the geographies of Hindu–Muslim violence, which were most visible and intense in 1991–2 in Ayodhya following the destruction of the Babri mosque by Hindu right activists (see van de Veer 1994) and in the Mumbai riots that followed in 1993 (Hansen 1999, 2001). Further wide scale violence in Gujarat in 2002 resulted in 100,000 displaced, 1000 Muslims injured and claims by human rights groups concerning the complicity of the state government in anti-Muslim violence (Varadarajan 2002; Engineer 2003; Lobo and Das 2006).

Following these events, considerable academic labor has been devoted to the task of deconstructing this apparent propensity towards violence in India, and the conditions and causalities underpinning such events. The debate has been typically characterized by positions that conceptualize "ethnic" violence as the normative narrative of inter-community relations, as an inevitable product of capitalist modernity and as the failed rationale of Indian secularism (Nandy 1995; Engineer 1995; Tambiah 1995; Kakar 1996).

In spite of major high profile incidents of violence and political tension between Hindus and Muslims, and the dominance of academic inquiry into what explains violence over recent decades, everyday life in much of urban and rural north India is *not* characterized by perpetual inter-community violence (Varshney 2001, 2002; Varma 2005; Tully 2007). Recognizing this, some political scientists have shifted their gaze away from such large-scale ideologies towards accounts that seek to uncover relational dynamics within civil society and political arenas and to open up space for thinking about why violence does not happen, as well as why it does (Varshney 2002; Brass 2003; Wilkinson 2004). This book develops that work by viewing peace as a process that is constituted within and reproduced through civil society and the everyday, and as shaped by individual agencies and local contexts. But, it also turns attention to the ways in which peace as a process is political and constituted through uneven geographies of power. It thereby seeks to encompass a view of the ways in which violence and disorder become attached and absorbed into the ordinary (Das 2007), and underpin the reproduction of everyday peace.

On Peace and Peaceful Sociality

So how might we start to understand complex, everyday experiences of peace? To examine this question I evoke a range of studies on peace and peaceful encounters and draw on work from peace studies and international relations, anthropology and sociology, before turning to consider more recent attention on peace in geography. The study of peace and war has been a central component in the discipline of international relations (Kant 1795/1991; Knutsen 1997), but it was not until the 1950s that "peace" became a discipline in its own right. At this time the USA and UK were grappling with questions about the causes and consequences of war, and peace research developed at the intersection of peace activism and modern social science (Gleditsch et al. 2013). Johan Galtung (1969, 1996) was one of the original founders of "peace research" and continues to be a preeminent theorist on peace. His work has been instrumental in making the case for peace as more than just an absence of violence. To counter this negative construction of peace, Galtung introduces the idea of "positive peace" through which he depicts a more full-bodied account of peace. This concerns not just the absence of violence, but also "the integration of human society," as constitutive of collaborative and supportive relationships and the presence of justice (1996).

An important aspect of Galtung's work on peace is the attention he dedicates to conceptualizing violence. He developed the idea of "structural violence" to describe situations of "negative peace" that have violent or unjust consequences and "originating violence" as an oppressive social condition that preserves the interests of the elites over the needs of disposed and marginalized populations (Galtung 1969). Galtung's work has been critiqued for, amongst other things, its black and white distinction between "negative peace" and "positive peace" (Boulding 1977). Boulding (1970) attempts to blur these boundaries in his discussions on "unstable" and "stable peace" and the recognition that peace has many different phases within it, which "may encompass greater or lesser justice, oppression, competence, enrichment, impoverishment, and so on" (Boulding 1988, p. 2). Yet, whilst the work of both Galtung and Boulding are useful prompts for highlighting the different forms and types of violence *and* concurrently peace, they are primarily concerned with the relationships between nation states and Western ideologies and represent normative constructions of peace.

The tone and focus of peace studies and its relationship to violence has shifted in line with the transformations from interstate to intrastate violence and the rise of "new war" cultures (Kaldor 1999). Responding to the rise of intrastate violence peace scholars have increasingly recognized the need to act in the midst of violent conflict "to ameliorate its consequences and prevent its recurrence" (Cortright 2008, p. 5). This has been witnessed in the rise of UN peacekeeping operations and peace-building missions in recent decades in the Congo, Sudan, Northern Ireland, the former Yugoslavia and Iraq amongst other places. As work on post-conflict peace-building has demonstrated, peace is a process. It is not a stage in time or an absolute endpoint but "a dynamic social construct" that demands input from peacekeepers, civil society and state actors (Lederach 1997, p. 20; Cortright 2008). Yet, it is interesting to observe the ongoing tension and debates within peace studies concerning the discipline's preponderance to focus on wars, with less attention directed towards understanding broader practices of violence, injustice, oppression and exploitation in society (Carroll 1972; Johansen 2006). And in recent years the studies that address peace more explicitly, do so almost exclusively through the lens of "liberal peace" (Gleditsch 2014).

Meanwhile the activist realms of peace studies have been connected with non-violent movements for social change: from the acts of civil disobedience by American Quakers during the American war of Independence to Gandhi's stance of non-cooperation under colonial India and the use of nonviolence as a political tool or technique by Gene Sharp (Holmes and Gan 2005; Thompson 2006). Put simply, nonviolence means abstaining from the use of physical force to achieve an aim; as an ethical philosophy it contends that moral behavior excludes the use of violence and as a practice it can be a strategy for social change, an act of resistance and the working towards peace. In contrast to "nonviolence," the notion of peace and associated ideas of pacifism, are sometimes regarded as apolitical and ineffective.

Despite importantly highlighting the nature of peace as an ongoing process that is constituted by different forms of violence rather than their absence, peace studies scholarship remains far better at proposing ideas about what peace *should* look like and who *should* do it, rather than examining what peace *does* look like. As an idealistic construct, Galtung's thesis for example, is less conceptually open to contrasting cultural productions of peace, the intimate and messy relationship between violence and peace and the ways in which hostile and antagonistic relationships are often also a part of peace. Peace studies fully recognizes that peace is a contested term which constitutes contested realities (Cortright 2008), yet there are calls within the discipline to focus more directly on peace, and to understand the contested and complex realities of peace from the "bottom up" (Jutila et al. 2008; Richmond 2008; Gleditsch 2014).

This book therefore takes up that invitation to explore the grounded realities of peace and empirically engage with peace as a process which is always complexly and intricately intertwined with forms of violence. The challenge therefore, in attempting to understand peace is to also expose the conflicts and injustices that pass as "putative peace" (Roy 2004, p.15), to expose the violence of peace.

The task of understanding and exposing the uneven politics of power that shape post-conflict settings and their representations has been taken up by International Relations scholars. In particular, they have highlighted the role of the discipline itself in perpetuating global governance norms through the uncritical language of "liberal peace" which, some argue, serves to conceal the underlying inequities and inequalities that perpetuate conflict (Duffield 2002; Pugh 2004; Pugh 2011 et al.). The term "liberal peace" is designed to capture the belief that liberal democratic states are inherently peaceful and serve as the ultimate model for so-called "failed states." Rather than transform systematic inadequacies of the neoliberal order, peacekeeping missions reproduce geopolitical structures and power relations which maintain patterns of marginalization (Higate and Henry 2009, p. 13, see also Richmond 2008). Oliver Richmond argues that "liberal peace" represents a form of "orientalism." He contends that liberal peace is dependent on elite actors "that know peace to create peace for those that do not"; peace is therefore a state which enlightened and rational actors impose on others. The notion that peace is created in a particular image, which both includes and excludes specific groups, should be taken seriously. Richmond reflects on the intimacy of peace and war, both of which are states of being and methods of political change (2008, p. 14). The especial potency of peace however, is that it represents an apparently universal view of an ideal worth striving for and is discussed, interpreted, and referred to in a way that nearly always disguises the fact that it is essentially always contested (Richmond 2008, p. 5). In problematizing the idea that "liberal peace" is inherently "good," Richmond contends that attention should be directed towards the everyday in order to fully comprehend and incorporate the role of situated agencies and local conceptions of peace within peace-building programs. Richmond proposes the notion of "hybrid peace agendas," which both integrate the "bottom up" and "top down" approaches (Richmond 2009, pp. 324–34), however, little work has advanced empirical understandings of local practices and experiences of everyday peace.

An interesting approach to understanding the scalar complexity of peace, and how local and national narratives of peace interact, is reflected in the research of John Heathershaw (2008) on peace building in Tajikistan. He highlights how the circulation of multiple narratives in the (re)production of peace find expression across different scales. He distinguishes between the international rhetoric of "peace building," the elite notion of "*mirostroitelstvo*" (Russian: peace building) and the popular "*tinji*" (Tajik: wellness/peacefulness) and suggests that different actors differentially draw upon these narratives in their everyday and institutional work towards maintaining peace (Heathershaw 2008, pp. 219–23). But, the focus on the narratives rather than the actors involved within these different spheres reveals little about what local agencies actually look like. My account is enriched by this recent intellectual shift in international relations towards a more critical perspective of the normative power of peace discourses and a greater appreciation of how peace policies are experienced and interpreted on the ground. However, international relations has so far been less effective at empirically illustrating the interconnections between the geopolitical and the everyday, and the situated interplay between narrative and agency in reproducing peace. By advancing an empirical approach to everyday peace, this book develops a crucial perspective on the role of legitimacy (and contingency) in shaping the contours of everyday peace. And, it contributes to a neglected area of investigation on the practice and experience of peace in "undramatic" contexts, away from postwar zones and away from international blueprints for peace (see also McConnell et al. 2014).

Examining the nature of peaceful sociality in "undramatic" contexts, has been the preserve of a few anthropologists concerned with "peaceful peoples" and "peaceful societies" (Howell and Willis 1989; Sponsel and Gregor 1994), and others concerned with the intimate relationship between violence and nonviolence in everyday contexts (Scheper-Hughes 1993; Ring 2006; Das 2007). These insights prove hugely instructive in three ways for thinking critically about peace. First, they draw attention to the importance of rich empirical understandings in illuminating forms of "microlabor" and ongoing effort articulated by different agencies in maintaining everyday peaceful life, and elucidate the uneven (re)production of power in particular places. Second they show how peace and violence are entangled in complex ways. The work of Veena Das (2007) is particularly instructive for illustrating how histories of emotional and physical violence insidiously inform seemingly ordinary life worlds long after the tangible realities of violence. Third, peace is not contingent on purely peaceful interactions and the successful resolution of tensions, but may also be constituted through the suspension of tensions and/or the articulation of relations that are less than peaceful. Laura Ring (2006) interprets everyday peace within the social spaces of an ethnically diverse Karachi apartment building. Her work importantly illustrates that peace is not an end-product, but rather is always in a state of becoming. Moreover, peace as a process does not necessarily involve or nurture the potential for transformation. To the contrary, the reproduction of peace may depend on maintaining uneven balances of power characteristic of the status quo. This account of peace builds on these studies which have been typically less successful at drawing linkages across scale and attending to the uneven realities of power.

At the heart of these studies on peace is the matter of how difference between people and ideas are negotiated and potentially transformed. The nature of encounter and interaction across difference has been the concern of sociologists, as well as anthropologists and geographers, interested in the social dynamics of multicultural cities. City life has long been recognized as "a being together of strangers" (Young, 1990, p. 240), whilst recent debate has focused on how the ways in which people live together across difference can inform and transform our relations with the "other" (Sennett 2011; Amin 2012).

Ash Amin argues that for encounters to be transformative the bringing together of people in particular places matters. He contends that "micropublics" such as the workplace, schools, colleges, youth centers, sports clubs and other spaces of association engender degrees of interdependence which demand dialogue and "prosaic negotiations" (Amin 2002, p. 14). Through these practices contextual identities and experiences may transcend notions of ethnic difference. Amin is optimistic that such sustained social contact within shared environments will offer the potential to destabilize dominant notions of difference with longer-term, progressive implications. This perspective resonates with literature in South Asian studies which has emphasized the role of particular settings in underpinning inter-community and intercaste relations, including civic publics (Varshney 2002), the workplace (Chandhoke 2009), tea stalls (Jeffrey 2010) and leisure pursuits (Kumar 1988). This account engages with these studies and extends Amin's concern with "micropublics" through an examination of economic, civic and political spheres in an Indian city. But, it also goes further to illuminate the ways in which matters of ethnic difference are entrenched and more often aligned with patterns of social and material inequality, and how "micropublics" may be reconfigured at different times, themselves located within shifting cultural political economies.

Yet, the nature of the encounter itself is also significant for inspiring acts of tolerance and respect, as well as shifts in previously held prejudices. In reality, it seems that coexistence proves contradictory and may involve both solidarities and disconnections between groups and individuals. For example, Humphries et al. (2008) have shown how the city of Bukhara was projected as a site of harmonious coexistence situated within global networks of connection whilst its residents simultaneously emphasized the nature of their boundaries and reserved intimate spaces. And, in Egypt, Bayat (2008) has documented how Coptic Christians and Muslims differentially imagine and articulate inclusive inter-communal connectedness and exclusive communal identity through the course of time and space.

Others are less optimistic about the potential for encounters within public spaces to challenge or destabilize the kind of inequalities and prejudices that underpin notions of majority–minority difference. Acts of tolerance that appear to structure exchanges may in fact obscure and thereby facilitate implicit power relations, where to tolerate someone else is an act of power; to be tolerated is an acceptance of weakness (Walzer 1997, p. 52). As Valentine (2008) has argued, positive encounters with individuals from minority groups do not necessarily alter people's attitudes towards groups as a whole and for the better, at least not with the same speed and permanence as negative encounters do. Others suggest that spatial proximity may actually inspire defensiveness and the bounding of identities and communities (Young 1990) rather than generate connections.

Yet, in order for people of apparent difference to live together, it is not always possible or necessary that friendships inevitably develop for peaceful sociality to be sustained. Frederick Bailey's return to his ethnography conducted in the 1950s in the east Indian state of Orissa is especially illuminating. Bailey (1996) argues that the village of Bisipara was cut through with difference; villagers saw each other quite literally as different breeds and arranged these breeds in hierarchy of worthiness. However, inhabitants adopted a "commonsensical pragmatism" about the extent to which such prejudices dominated their everyday lives. Accordingly, they articulated strategies to negotiate possible tensions, including practices of ritualized politeness, and attention to the etiquette of status. Whilst notions of difference were firmly moralistic, they were simultaneously domesticated and kept under control. In this setting Bailey argues that the "civility of indifference" prevented the potential for tensions and conflict to escalate into violence. The notion of "indifference" as a productive way of being with others resonates with the theories of situated cosmopolitanism (Sennett 1994; Sandercock 1998; Donald 1999; Amin 2002; Tonkiss 2003), and challenges the argument that intercommunity relations need to be transformative in order for peaceful coexistence to prevail. Bringing scholarship on geographies of encounter into dialogue with conceptualizations of peace as an ongoing and precarious process helps to illuminate the ways that practices and experiences of encounter inform peaceful realities. Encounters generate very different kinds of outcomes between individuals and groups, and across sites and scales. This book attempts to understand, in part, how the uncertain outcome of encounters contributes to the contingent reality of everyday peace.

In view of these wide-ranging disciplinary studies, the framework of everyday peace offers important analytical purchase. It foregrounds particular modes of prosaic and "less than violent" (Darling 2014) interaction concerning, coexistence, friendship, tolerance and indifference, and enables a wider perspective of how these practices are differentially constitutive of peace. Thinking about everyday peace opens up the possibility for interpreting how contrasting experiences of coexistence, tolerance and indifference, for example, may conjoin and constitute the reproduction of peace for different people, realized in and through different spatial practice and narrative.

Geographies of Peace

This account of everyday peace contributes to and contrasts with emergent work in geography that is explicitly orientated towards understanding and problematizing "geographies of peace" (see McConnell et al. 2014), antiviolence, (see Loyd 2012) nonviolence (Woon 2014) and "non-killing" (see Inwood and Tyner 2011a, 2011b). The history of geography's engagement with the concept of peace has been well documented and shows a discernible shift in the discipline from actively making war to averting war (see Mamadouh 2005; McConnell et al. 2014). This book responds to calls by Gerry Kearns for a "progressive geopolitics" (2009) which privileges the view that humanity is not predisposed towards enmity and calls for better understanding of the constructive nature of geopolitical relations outside of the nation state. Developing this geopolitics of peace Nick Megoran's "pacific geopolitics" is orientated towards understanding "how ways of thinking geographically about world politics can promote peaceful and mutually enriching coexistence"

(2010, 2011: 185). Nick Megoran's (2010, 2011) research is illustrated by the strikingly transformative effect of Crusade apologies of Christian missionaries and is productive for considering the different ways in which people actively orientate themselves towards one another in the (re)production of peace. However, this account diverges from Megoran's approach in two ways. First, it is not motivated and framed by a Christian morality towards peace, and, secondly, the account portrayed is notably less optimistic about the inherent potential for transformation in the lives of those who practice peace. The arguments set out in this book adopt a more cautious approach that grounds peace in a specific context and, informed by a subaltern view of peace, reveals the uneven geographies of power through which peace comes to be practiced and experienced.

The production and reproduction of peace is inherently spatial, where space is both the context for and the product of peace, whilst peace is also shaped by the spaces in which it is made (Koopman 2011b). Here, it is useful to evoke the work of Massey (2005), who conceptualizes space as the unfolding of social interaction. She argues that to think about space is to think about the social dimension, the contemporaneous coexistence of others. As the sphere of the continuous production and reconfiguration of heterogeneity in all of its forms – diversity, subordination, conflicting interests and connection – space is therefore relational. It is the dimension which poses the question of the social and therefore of the political. More specifically, Massey is concerned with "the *character* of relations and their social and political implications" (emphasis mine 2005, p. 100). This sensitivity to the spatial and social implications of peace reflects concern with orientations towards tolerance, trust and hospitality, but it also underpins geographical approaches which show how space is instrumentalized to control bodies within conflict and post-conflict zones, as well as within "peaceful" societies. The deployment of state and/or unilateral force may actively if not also aggressively secure spaces of peace through practices of "peacekeeping" (Grundy-Warr 1994; Higate and Henry 2009), coercive technologies of government (Lyon 2004; Moran 2008; Coaffee and Murakami Wood 2008), methods of containment (Alatout 2009) and the enforced exclusion of apparently violent bodies (Beckett and Herbert 2010). Actions towards keeping the peace concern material practices in space, made manifest in the form of the wall in the West Bank (Cohen 2006; Pallister-Wilkins 2011), the Belfast "peace lines" or "peace walls" (Nagle and Clancy 2010, pp. 79–80) or the conceptual demarcation of "no go" areas (Peake and Kobayashi 2002). Through the securitizing and narrating of space, places are relationally constituted as "safe cities" (Hyndman 2003) and "zones of peace" (McConnell 2014).

The spatially contingent and strategic nature of peace is exemplified by the fascinating work of Sara Koopman on protective accompaniment and transnational solidarity. She shows how the positioning of particular bodies within space and alongside particular others creates useful local sites of security (2011a, 2011b, 2014). In Columbia, protective accompaniment "makes space for peace" by putting bodies that are less at risk next to bodies that are under threat in order to enhance mobility and enable their struggles to build peace and justice in the midst of conflict. This is revealing of the way that peace can be embodied, and space manipulated by different actors as a form of strategy.

How peace interacts with different people and places is intimately related to peace discourses; the ways in which peace is imagined by diverse actors and circulated through networks of knowledge and experience. Like their contemporaries in international relations, geographers have been critical of the ways in which the liberal peace has been conceived of, implemented and justified in different contexts (Jeffrey 2007; Springer 2011). They have also documented the situated contradictions and contestations of the liberal peace agenda, as Patricia Daley (2008, 2014) illustrates in her work on postconflict Central Africa, where the dissonance between liberal peace programs and local cultural contexts has inadvertently reinforced gender inequalities and local insecurities. Others have examined how public policy processes seek to transform identities and relations in postconflict polities (Graham and Nash 2006) and how cultural policymaking does constructively inform more everyday peaceful realities (Mitchell 2011). But discourses of peace do not only originate with the state and international actors. Geography scholarship has shown how an international language of human rights is appropriated and reworked by NGOs to build local, domestic practices of peace, for example in Uganda (Laliberte 2014), and how discourses of nonviolence are reworked by different actors for political strategy (McConnell 2014). This book promotes a more grounded interpretation of peace to develop an understanding of the ways that local, embedded and vernacular expressions of peace emanate within particular places and resonate differently for people.

To understand how discourses are circulated and deployed towards the generation of peaceful realities, it is necessary to understand the role of agency, how actors are located in particular contexts, how their actions acquire legitimacy, and why it is that constructive agency may be articulated in some times and spaces and destructive agency in others. For example Chih Yuan Woon's (2015) move to "people" peaceful geographies serves to destabilize perceptions of the military in the Philippines as warmongers and to illuminate the place of military personalities and their relations with civil society organizations that work to maintain peace. Other geographers have highlighted the complex and contingent role of religious brokers in facilitating peace (Johnson 2012), the ways in which charismatic celebrities come to embody peaceful international relations (Megoran 2014), and how political leaders may become synonymous with peaceful imaginaries and agendas (McConnell 2014).

The focus on agency has been hugely productive for understanding the everyday and embodied dimensions of peace; nonetheless it tends to privilege the role of elite actors rather than ordinary citizens interpreting and orchestrating peace on the ground in both mundane and extraordinary circumstances.

My account is empathetic to the principle held by "non killing" geographies: that as geographers we have a responsibility for destabilizing rather than reproducing the boundary between war and peace and challenging taken-for-granted norms concerning cultures of war and practices of killing. The underlying rationale is that pursuing these goals will enable geographers to address social and economic inequality through the use of nonviolent practices, work toward the promotion of a lasting peace, and incorporate those practices in the classroom through a pro-peace pedagogy. However, my interpretation of "non killing" scholarship is that it largely aims to destabilize the war–peace binary by problematizing *war*, rather than problematizing *peace* as the emerging scholarship on "geographies of peace" (McConnell et al. 2014) strives to do. Despite the significant and exciting rise in attention towards peace by geographers there remains a conspicuous lack of commitment to actually understanding what peace looks like from a subaltern perspective (for exceptions see Koopman 2011); how it is constituted as a process within the spaces of the micropolitical; and how everyday peace activities are indicative of and entangled with patterns of power and "the political" across different scales. In order to address these concerns and show how peace, space, discourse and agency are connected, I turn to feminist geopolitics.

Feminist Geopolitics, Violence and "*Everyday* Peace?"

Critical geopolitics, and especially feminist geopolitics affords a useful framework through which to foster a grounded perspective on peace that is especially attentive to agency, power and the political in and through different sites and scales. Here, I identify four important approaches for conceptualizing everyday peace. To begin with, feminist geopolitics destabilizes the primacy afforded to geopolitical action and discourse, and shifts attention to the everyday and routine spaces of life. Feminist geopolitics has been expressly engaged with revealing the local, everyday and embodied manifestations of war across different sites (Dowler and Sharp 2001; Cowen 2008; Enloe 2010). Others have shown how citizenship and identity becomes reconfigured (Cowen and Gilbert 2008) and how war narratives take on gendered dimensions

(Dowler 2002). De-centering the role of geopolitical actors and the state in matters of violence and security opens up space for considering the "other kinds of securities" that are possible, that operate outside of the government and are otherwise "off the page" (Koopman 2011a). This is important for thinking about lived realities of peace, where Koopman challenges the geopolitical focus on peace practices (the peace treaties, peace accords, peace-building programs) by thinking about and working with "peace communities" and "protective accompaniers" in situ. Foregrounding embodied experiences privileges an understanding of the emotive infrastructures (Woon 2011, 2014) of war and peace, and how emotions such as fear (Pain 2009), danger (Megoran 2005) and feelings of security or insecurity (Dodds and Ingram 2009) differentially structure everyday war *and* peace settings where memories of violence linger and conflicts remain suspended. Second, as well as uncovering the violent politics of difference and how these are embodied and experienced, feminist geopolitics has examined the connections and the conditions under which alliances are formed across scales (Katz 2004; Koopman 2008). Underlying these approaches is a concern that "the political is not just about differences – either between people or between perspectives; it is also about the webs of power and social relationships that are the basis of connections" (Staeheli and Kofman 2004, p. 6). I would also suggest that sometimes the presence of differences may underscore the imperative to build and sustain solidarities and connections.

In these respects, a feminist geopolitical approach to peace inspires a grounded understanding of how the geopolitical practices of peace are produced through the everyday, and are made manifest at the level of the body, household and community, city, region and nation. It therefore enables a critical perspective of the ways in which peace is constituted both through narrative and practice, and intertwined with processes across scale. For instance, in the Indian context this prompts an examination of the ways in which Indian state policies of secularism inform practice of tolerance and coexistence "on the ground" as well as the ways in which geopolitical action and discourse interacts with everyday life. And, it helps to show how practices of connection and disconnection are differentially engendered, made visible through sartorial choices, deportment, local social and economic alliances, institutional discourses, urban life and media rhetoric.

Third, of particular importance to this account is how India's Muslims experience everyday peace in the margins of society. Feminist geopolitics grounds politics in practice and in place in a way that makes the experiences of the disenfranchised more visible (Koopman 2011a). Such exposure necessarily attends to matters of subaltern agency (Sharp 2011), which may be situated within particular histories of violence and nonviolence (Dowler and

Sharp 2001; Abu-Lughod 2008[1990]). By examining individual and group practice, this book contributes to critical scholarship that recognizes that agency is often more than simply resistance. Katz's reworking of resistance through her research on youth in Howa, Sudan and New York, USA provides us with some useful tools for thinking about resistance. She differentiates between "resistance" that involves oppositional consciousness and achieves emancipatory changes, acts of "reworking" that modify the structure but not the distribution of power relations and "resilience" as the capacities that enable people to survive but do little in the way of transforming the conditions that make their survival so difficult. Contextualized accounts of agency in different sites are important because "they can attend to its variations, including its limits, its structuring context and its uneven impact, rather than simply its autonomous existence" (Sparke 2008, p. 424; see also see McNay 2000, 2004).

Expressions of resilience (Katz 2004), the "politics of patience" (Appadurai 2002), practices of waiting (Jeffrey 2010) and strategies for "dealing with and getting by" (Sharp 2011), therefore all represent significant modes of everyday agency shaped and enabled through networks of power and which may inspire more subtle forms of transformation (Katz 2004). Adopting this approach "encourages conceptualizations of agency not simply as a synonym for resistance to relations of domination, but as a capacity for action that specific relations of *subordination* create and enable" (Mahmood 2005, p. 17). This approach is important for understanding the ways in which practices of peace interact with the *status quo*, and may or may not be transformative.

Fourth, feminist geopolitics has recognized and problematized the ways in which violence and nonviolence, war and peace, are intertwined, and addresses the unequal relations between people on the basis of their real or perceived difference. Sara Koopman makes the case that whilst feminist geopolitics explicitly "eschews violence as a legitimate means to political ends" (Koopman 2011a, p. 277) it has done little to offer up alternative solutions. In response, Koopman (2008) proposes "alter geopolitics" to reflect the ways in which grassroots movements are not only pushing back against the global hegemonic powers and spaces of (in)security, but are also cultivating new forms of nonviolent practices and securities in autonomous, non-state spaces. Jenna Loyd deploys the concept of "antiviolence" to counter the dominant liberal state narratives of domestic peace that juxtapose war abroad. She contends that feminist approaches are vital for "understanding interlocking relations and spaces of violence, who is most vulnerable to harm, and how people work for justice and peace that can transform the conditions in which they live" (Loyd 2012, p. 486).

To understand how spaces of violence and peace are intertwined it is important to recognize how different forms of violence (see Brubaker and

Laitin 1998) intersect and reconfigure local intercommunity experiences and subjectivities across different scales. Violence between Hindus and Muslims in India has been typically characterized as "collective violence" which takes the form of "riots." This conveys a sense of "spontaneous" or "instinctive" eruptions of violence which may or may not be attached to certain primordial identities or communities and which take place in the public arena, enacted by citizens rather than agents of the state. In reality, collective violence is rarely "impulsive" and does not represent an aberration of normal life, but construing it as such serves to depoliticize and dehumanize violence by detaching it from its socioeconomic and political contexts as well as its longer-lasting legacies (Hansen 2008, and for historical perspective see Bayly 1983; Pandey 1990).

Tyner and Inwood (2014) have cautioned against a tendency within social sciences to treat violence as fetish such that violence becomes abstracted from its concrete forms and affects. They go beyond Galtung's distinction between "structural violence" and "direct violence," to propose a dialectical approach that is simultaneously concerned with 'real' concrete forms of violence and their abstracted affects. Allen Feldman's (1991) excavation of political violence in Northern Ireland usefully shows how violence sits in sites, bodies and spaces; the state, the city, the neighborhood, the jail, the jail cell, the gunmen, the body of the hunger striker. In particular, he emphasizes how political violence achieves and works through the "instrumental staging and commodification of the body" (1991, loc.135 of 4562). This view of violence echoes Scheper-Hughes and Bourgois's (2004) conceptualization of violence as a social process that has a "very human face" which gives violence its meaning and force. Indeed, understanding why people kill and do violence involves seeing social worlds as political and historical products. But there are deep ambiguities in recognizing violence, and what counts as violence is a matter of perspective that is often best understood through an empirically grounded approach.

In her ethnographic study of survivors of riots in India, Veena Das (2007) shows how violent events become absorbed into everyday life, and are always attached to the ordinary, not just as memory and experience, but also as fear of the potential. Similarly, Gyanendra Pandey (2005) proposes the concept of "routine violence" to access the way in which violence is routinely folded into the everyday and ordinary. He argues that violence against India's minorities is palpable, not just as violent events but also in the exclusionary nationalist discourses which underpin constructions of community and belonging, and concurrently the production of the nation's "marked" and "unmarked" citizens. Recognizing how situated experiences of material and rhetorical violence shape everyday peaceful realities is therefore an important dimension of this book.

Unlike the studies by Das (2007) and Feldman (1991) this book does not take the immediate presence or "aftermath" of violence as the subject for study. Rather it focuses on a city which is locally and nationally represented as a site of "peace" between Hindus and Muslims yet situated within wider spatial and temporal experiences of violence. By examining the presence of peace, the book does not eschew forms of violence but instead seeks to interpret how peace is negotiated and framed in dialogue with routine, physical and potential violence. Approaching this problem from the perspective of the *everyday* focuses attention on human agency and the micro-politics of interactions to counter conceptions of peace that are disembodied and romanticized. It offers an insight into how ordinary people negotiate spaces, sites and bodies of violence and peace, justice and injustice and it reveals how practices of peace are contested and negotiated in the everyday. It highlights the scalar nature of narrative and practice, and how these intersect and inform events and experiences that cohere in particular places and have particular implications for the construction of everyday peace (on the everyday see Rigg 2007).

Geographers have typically engaged with the everyday as a means to contrast other categories deemed extraordinary or spectacular such as the carnival (Lewis and Pile 1996) and parades (Brickell 2000). In its role here as a residual category the meaning or existence of "the everyday" is rendered uncertain. Froystad (2005) argues that the apparent ambiguity around the concept of the everyday constitutes its most valuable function, in its capacity to provide the milieu for the main subject of research. More importantly, it also acts as a repository for phenomena, particularly those not naturally under the spotlight of research categories (2005: 18). Everyday life can also refer to lived practice, both bodily experiences and practical knowledges that constitute social experience and intersubjectivity (de Certeau 1984; Lefebvre 1991; Gardiner 2000).

Focusing on everyday life and the everyday is therefore a tactical maneuver to question the peaceful practices and narratives that take on the appearance of being taken-for-granted and routine (Felski 2000; Dowler 2002). It is a way of uncovering what lies beneath peace, examining how peace is held together, and asking by whom, with what intentions, and for what purposes, is peace remade. The trouble with interpreting everyday experiences through everyday narratives is recognizing when these serve as a "thin simplification" of reality (Scott 1990, p. 319). Yet, sometimes the discourses used to construct everyday life are simultaneously working to naturalize and normalize a context that is in reality far more precarious and fragile.

Sensitive to this, the book follows the concern of Navaro-Yashin (2003) in asking not merely what everyday life is about, but also attempting to draw out the disaster and disorder that underlies the seeming pretense to normality, by working against rather than simply depicting the normalizing discourses of informants. The uncertain juxtaposition of "everyday" and "peace" in the book's title is therefore intended to convey something of the precarious and incomplete reality of peace in north India. It gestures to the intimate relationship between peace, violence and the everyday. Further, its purpose is to agitate on the matter that narratives of peace need to be questioned, for they are often political, and their reproduction is contingent on concealing or coopting uneven geographies of power.

Citizenship as Inclusion and the Scalar Politics of Peace

I propose that the practice and articulation of citizenship is an important concept for interpreting the scalar politics of everyday peace and the role of agency and practice from within the margins. Citizenship facilitates thinking about vertical relations with state and geopolitical practices and imaginations as well as horizontal relations – the connections *and* disconnections – that exist within society. This book contributes to understandings of citizenship by showing how citizenship is concerned with structures of inclusion and exclusion, and strategically negotiated by people within different urban settings in ways which draw on and are orientated towards reproducing "everyday peace." My contention is that from the margins, these strategies and struggles for everyday peace respond to and are orientated towards realizing practices of tolerance, trust, civility, hospitality and justice.

Increasingly, scholarship conceives of citizenship as more than a legal or place-bound identity, arguing instead that it might be more appropriately understood as a set of discourses and practices that are rendered unevenly across asymmetrical social groups and contexts (Benhabib et al. 2007; Staeheli 2008 and Secor 2004; McConnell 2013). Through routinized habits, practices, customs and norms, a sense of meaning, belonging and identity may come to inform lived experiences of justice and rights (Lister 1997; Werbner and Yuval-Davis 1999), which may be interpreted in a comparative sense rather than against a fixed ideal (see Sen 2009). As such, this book responds to calls from citizenship studies (Nyers 2007) and geography (Desforges et al. 2005; Barnett and Low 2004) to examine actually existing citizenship. That is, how citizenship operates as "lived experience" (Nyers 2007, p. 3).

The fundamental issues of our time may concern struggles over citizenship (Isin et al. 2008), so what does citizenship mean in India, how is it actually realized and currently theorized? For marginalized communities, there is more often a disconnect between citizenship as *status* bestowed through membership of a nation (for example Marshall 1950) and citizenship as it is *actually practiced* (see Holston 2008), articulated in and through the social (Isin 2008). In some postcolonial nations the extent to which members are citizens or subjects is a significant question, where the transition to democracy has been partial and the notion of citizenship differentially enacted (Mamdani 1996; Chatterjee 2004; Stepputat 2004; Holston 2008).

In their project to decenter the state and recast the social in citizenship, Isin et al. (2008) invert the teleological acccount of citizenship proposed by Marshall (1950). Where Marshall theorized that citizenship primarily concerned the granting of civil rights, then political and, finally, social, Isin et al. (2008) suggest that we should recognize citizenship as firstly social before it can be civil and political. This resonates with Naila Kabeer's (2007) contention that the potential to actually realize the rights endowed through political citizenship in postcolonial space is more often contingent on the constructive nature of horizontal relations within society. When considering the relationship between the social and citizenship it is worth reflecting upon the ways in which "the social" has come to bear upon the theoretical and actual realities of citizenship in the Indian context. From the outset, the Indian Constitution, unlike the American, *was* heavily invested in the social and especially in questions concerning social uplift, equality and social identity (Mehta 2010). The government of India's deep concern with, and ultimate involvement in, the social is a product of the context in which Constitutent Assembly debates took place between 1947 and 1950 (Tejani 2008, see also Bajpai 2002). More specifically, it relates to the inherent and inherited diversity of India's society and the ways that this would inexorably inform the simultaneous project of nation building and unity. Given the fissiparious tendency of India's social domain, it was thus decided that state power and the political afforded the capacity to unify the nation, through a form of liberal citizenship underpinned by the ethics of secularism.

In Constitutional terms, Muslims are formally and fully citizens of India (see Roy 2011; Jayal 2013). However, political transformations in recent decades, most notably the rise of Hindu nationalist forces, have contributed to the creation of a social and political consensus that citizens enter the public sphere preconstituted by their identities, and that mediation by ascriptively defined communities constitutes the legitimate form of interaction between the citizen and the state (Jayal 2011, p. 145; see also Roy 2010). It is clear that the state alone could not guarantee citizenship where social and political forces also play an influential role in shaping the normative meaning of who is "included" and who is "excluded" from the national political community. Indeed, "political rationalities rarely find their exact mirror image in citizenship practices or politics. Their imprint on both the subject and the collective is always mediated by institutional contraints, different forms of resistance and, indeed the accidents of history" (Brodie 2008, p. 23).

Even whilst Indian citizenship privileged the importance of the social in structuring encounters between society and the state, the possibility of individuals fully enacting their citizenship has proven to be fundamentally contingent on experiences of, and positions within, the social (Isin 2008). In reality, groups are socially constructed, spaces of exclusion and inclusion are determined through social processes, and unequal relations are centrally located (Isin et al. 2008, p. 11). Invariably, the dominant population produces and reproduces normative citizenship narratives, which inevitably rest on the exclusion of others (Isin 2002; Chandhoke 2005), may shift for political gains, and often transcend the nation or state. In India today, public life is arguably dominated by "banal, everyday forms of Hindu nationalism" (Nanda 2009, p. 140; and Mathur 2010), which insipidly exclude India's Muslim as well as other religious minorities from fully participating in social and political life.

Isin (2002) asserts that exclusion is integral to struggles for inclusion and that the logics of alterity are inherent to constructions of citizenship, which are fundamentally about our relations with immediate others (2002, pp. 22, 25, 29). Ó Tuathail (2005) furthers this line of argument to emphasize the significance of the geopolitical in constructing notions of the Other and reconfiguring local practices of citizenship. But, reflecting on her own research as well as that of others, Staeheli argues that citizenship would be more aptly presented as "a quest for inclusion" (2005, p. 350). Where "inclusion" represents a distant, if not unobtainable reality, the focus of study turns to process, struggle and contestation as a means to recognize and question the inequality and exclusion that inhibits the full democratic potential. I find this argument particularly compelling, especially in a global context when the vast majority of the world's inhabitants are struggling to be included in society. I propose that the struggle for inclusion is in itself a practice of citizenship that engages with situated questions of justice and injustice and eschews universal constructions of citizenship to instead articulate forms of mundane practice that communicate solidarity and trust.

Of interest here is how this more inclusive and hopeful account of citizenship informs everyday actions and struggles to reproduce everyday peace. The book shows how everyday peace and citizenship interact in this particular geographical place where actors variously engage with peaceful concepts as expressed through policy, narrative and practice in order to articulate their citizenship from the margins, contest exclusionary and sometimes violent imaginaries and thereby struggle for inclusion. The manner and extent to which citizenship is actually realized by Muslims in the context of everyday peace also informs an understanding about the transformative potential of everyday peace for a particular subaltern group.

Political geographers have shown how citizenship and space are closely entangled (Smith 1989; Dowler and Sharp 2001; Miraftab and Wills 2005; Pain and Smith 20008; Dikeç 2009). In her study of Kurdish women in Istanbul Anna Secor (2004) views citizenship as a spatial "strategy" that is negotiated through the hiding and unhiding of different identities in everyday life. And, both Secor (2004) and Staeheli (2008) have shown with respect to communities how visibility may be an important strategy for empowering communities and facilitating citizenship, but at other times may undermine partial practices of citizenship that are realized.

As the work of both Kabeer (2002) and Isin (2002 and with Nielsen 2008) illuminates, underpinning claims to citizenship is often the question of justice or injustice as it is actually experienced and interpreted "on the ground." In line with shifts in geographical thinking towards a socialized, situated understanding of justice (see Barnett 2011; Fincher and Iveson 2011), the book finds great value in the arguments presented by Amartya Sen (2009). He proposes that we need to understand justice not in terms of the normative frameworks with fixed notions of maximum justice, but to engage empirically and relationally with justice as it is actually realized along a continuum. Linking conceptualizations of justice and everyday peace elucidates agency and the uneven geographies of power that constitute and are constitutive of peace. As such, the book problematizes the idea that peace entails perfect justice, and theorizes the differential ways in which justice is conceded, suspended and struggled for in contrasting relational and scalar contexts. A theme running through the book, therefore, concerns the contingent relationship between visibility, justice, community and citizenship and the ways that this is differentially negotiated by Muslims in the context of realizing everyday peace.

As practices of citizenship and efforts to maintain peace have been located within the domain of civil society by some (Janoski 1998; Varshney 2002), this provokes questions about the nature of civil society in the Indian context, and its relation to the state and citizens. The traditional, "Western" notion of civil society constitutes the area of public space between the state and the individual citizen (or family), which makes connections between individuals, families or groups possible. Associations that emerge in this space take an organized and collective form which exist outside the direct control of the state (Held 1989, p. 6). Through these formal institutions disinterested individuals come together and engage in a dialogue about "the public good." A further assumption is that civil society consists of more or less voluntary associations outside the state. At this point however, the idea of civil society in Indian society departs from the Western-derived concept. As Elliott (2003, p. v) asks, does it include "only western style voluntary associations or the larger array of groups in the Indian social environment?" The allure of civil society in India, as indeed elsewhere, is rooted in a general disenchantment with the state (see Chandhoke 1995; Seligman 1992). In India, social movements, cultural assertions and affirmative action have all been linked with civil society, so that it has become identified with tradition and ascriptive bonds and not exclusively with voluntary associations (Gupta 1997). Furthermore, ethnic and religious associations can combine ascription and choice; for example, not all Hindus have to be members of a temple in a given town (see Varshney 2002).

While I employ the state and civil society as investigative tools within the book, it is important to problematize this distinction. Developed along the tradition of European anti-absolutist thinking, this delineation has the analytical disadvantage today of either regarding the domain of the civil as a depoliticized domain in contrast with the political domain of the state, or of blurring the distinction altogether by claiming that all civil institutions are political (Chatterjee 2001, pp. 171–2). Chatterjee finds it useful to keep the Hegelian notion of a distinct state and civil society. In light of the formation of mass political formations he finds it productive to think in terms of a field of practices mediating not between state and citizens in civil society, but between governmental agencies and population groups. Between those that govern and those who are governed. Unlike in "the West," Chatterjee argues that "[m]ost of the inhabitants of India are only tenuously, and even then ambiguously and contextually rights-bearing citizens in the sense imagined by the constitution" (2005, p. 83). Although people in this group are not regarded as members of civil society by the state, they are not positioned outside the reach of the state and the domain of politics. Instead, Chatterjee argues, they occupy the realm of "political society" which encompasses the relations between the governmental agencies and population groups that are targets of government policy.

In this reading of society-state relations, Chatterjee is pointing to alternative spaces in which citizenship comes to be realized for India's marginalized groups, spaces not in line with the formal constitutional contract. These are zones of "paralegal practices, opposed to the civic norms of proper citizenship" (2004, p. 128). Within these spaces also, Chatterjee senses the possibility for new, and often contextual and transitory, norms of fairness and justice in providing the welfare and developmental functions of government to large sections of the poor and underprivileged people (2004, p. 128). This book contributes to knowledge that at once challenges conventional readings of "civil society" and further nuances our understanding about the everyday manifestations of "political society," and how individual and group agencies are both informed by and shape this arena in often civil and peaceful ways (e.g. Jeffrey 2010; Mannathukkaren 2010; Gudavarthy 2012).

Muslim Geographies: Experience, Identity and Agency

In part, this book contributes to perspectives within geography and social sciences on Muslim experience, identity and agency in multicultural spaces. Such work came to prominence in the late 1980s and 1990s following the publication of Salman Rushdie's *Satanic Verses*, which raised questions for the UK and Europe about multiculturalism, liberalism and religious equality (Asad 1990, Modood 1994). In the aftermath of 11 September 2001 and the ensuing geopolitics associated with the "War on Terror" these questions gained renewed traction and importance, but this time with greater attention turned towards actually lived Muslim lives. A dominant objective has been to expose pervasive and often pernicious misconceptions about Muslim identity and agency as variously inward looking, conservative, aggressive and incompatible within Western multicultural societies (Dwyer 1999a, 1999b; Modood et al. 2006). Others have shown how Muslim experience is shaped in and through different scales, from the local to the geopolitical (Hopkins 2004, 2007), and examined how Islamic religious structures become symbolic and contested sites of identity articulation in multicultural cities (Naylor and Ryan 2002; Dunn 2005). Meanwhile, Tone Bringa's (1995) work in Bosnia is instructive for highlighting the types of corporeal and discursive strategies that Muslims employ in their everyday lives, in order to negotiate geographies of connection and disconnection.

My attention to Muslim lives builds on this commitment to challenge dominant understandings about Muslims as the "uncivil terrorist" or "helpless victim," and to develop more fine-grained understandings of Muslim lives and agencies as they are negotiated through different sites and scales of everyday life in multicultural India. At once subordinate and rendered invisible within the majority Hindu construction of India, India's Muslims have been paradoxically central subjects in the project of nation building and (re)defining India's imagined and physical borders. The provocative "Muslim question" – "can a Muslim be an Indian?" (Pandey 1999) – continues to resonate in some sections of Indian society and media. In many ways, the simultaneously ambiguous and contested status of India's Muslims undergirds the polarization of narratives around their membership in the contemporary Indian nation.

It should be kept in mind that far from being a unified, internally coherent group, India's Muslims represent a diversity of educational and employment backgrounds, regional and linguistic identities and cultural aspirations. Nonetheless, political and academic enterprises have contributed to a reification of India's Muslim "community." This has implications for imagining Muslims as citizens, and interpreting their engagement with citizenship as a process of both self and subject-making and how this is wrapped up with experiencing and realizing everyday peace.

Counterpoised to imaginations around the Muslim aggressor, dominant popular and academic discourses also construct Muslims as "helpless victims." Scholarship documenting the condition of India's Muslims since independence has importantly highlighted their economic marginalization (Hasan 1996), lack of political representation (Ansari 2006), struggles to see Urdu recognized as a national language (Abdullah 2002) and low levels of education and literacy (Alam and Raju 2007). Muslim women are regarded as doubly marginalized; as women and as Muslims they have unequal access to citizenship (Hasan and Menon 2004). A report by a high-level parliamentary committee chaired by Chief Justice Rajinder Sachar and presented to the government of India in 2006 determined that India's Muslims experience widespread discrimination and marginalization in all walks of life; in education, employment, in access to credit and political representation. And, Muslims are routinely victimized at the hands of the Indian police (Khan and Mittal 1984; Varadarajan 2002; Brass 2003; Engineer 2003).

Stories abound in the national and regional media about innocent Muslim males arrested on charges of terrorism, held indefinitely without official criminal cases filed against them, or worse, shot in the course of police encounters. The "Batla House encounter" involving the fatal shooting of two Muslim students by New Delhi police in 2008 provides one such high-profile incident. Discriminated against by the state and society, Muslims are portrayed as having little recourse to agency or opportunities for self-representation, whilst the poorest amongst them are portrayed as "people without history" (Seabrook and Siddiqui 2011). The conclusions often drawn are that Muslims chose to segregate themselves and look inwards to the Muslim community rather than participate within Indian society. In many ways they have become socially and imaginarily ghettoized.

These polarized and contradictory normative constructions of India's Muslims as "terrorist" or "helpless victim" function to depoliticize the dominant structures of power that mediate the ways in which Muslim experiences of state–society encounters are produced in and through the social. And, they conceal the possibility of another kind of Muslim citizen as actively committed to the secular good, pragmatic and resilient about the possibilities of asking questions about justice. Turning attention to Muslim identity, agency and experience in the context of questions about citizenship "from below" opens up the possibility for examining alternative perspectives about Muslim agency and intercommunity realities. This is where citizenship may be conceptualized as "the art of being with others, negotiating different situations and identities, and articulating ourselves as distinct yet similar to others in our everyday lives, and asking questions of justice" (Isin 2008, p. 7).

Read this way, what becomes interesting to explore is the process of negotiation and the kinds of strategies deployed by different Muslims to "get by" or "get on," and how these are informed by particular contexts and respond to and through different scalar practices. Based on their fieldwork in rural northeastern Uttar Pradesh in India, Jeffrey et al. (2008) show how local and geopolitical events shaped the strategies of young Muslim men as they sought education and employment opportunities. When looking for work outside Muslim circles or interacting with state officialdom and the police, young men actively negotiated and moderated their educated Islamic identities to limit or avoid negative feedback. On the whole, they tended to withdraw from, rather than assert themselves in political spheres, which functioned to reinforce their exclusion. The study usefully illustrates how cultural and political influences significantly shaped the types of strategies employed by Muslim men.

Similarly, in his study of the Jamaat e-Islami Hind in contemporary India, Ahmad (2009) is sensitive to how Indian Muslim actions are situated within India's particular cultural political economy. He looks at how this Islamist organization has encountered, engaged with and undergone transformation within India's secular democracy. He argues that the Jamaat's evolutionary trajectory reflects a strong engagement with Indian secularism which has informed the party's progressive democratization. And, he contends that radical Islamist actions, notably by Students Islamic Movement of India (SIMI), are best understood within the context of Indian Muslims' discrimination and exclusion from mainstream society. Meanwhile Rubina Jasani (2008) has documented the constructive agencies of Islamic relief organizations in providing spaces of care after the Gujarat earthquake where the state government failed to provide humanitarian relief.

These studies document contrasting situations in which Muslim individuals and groups differentially engaged with the state and society under shifting circumstances, expressing different forms of agency from withdrawal, to resistance and pragmatism. Examining everyday peace mainly through the voices and experiences of Muslim informants disrupts constructs of Muslims as the local and global subaltern of contemporary times with little recourse to agency. Rather, it offers a particular perspective on Muslim action in public urban life as situated within particular histories of violence and nonviolence as well as state practices and social realities. It shows how Muslim individuals and groups articulate citizenship in diverse and creative ways, differentially engaging with the state and society in ways that often forge spaces of hope of connection (Mills 2009; Phillips 2009, 2010; Phillips and Iqbal 2009; Jones 2010) and sometimes engender the potential for transformation.

This is not a book about religious thought and action, but it does appreciate the lived and practiced nature of Islam, and as such contributes to studies examining the ways in which religion constitutes an important social identity which is differentially constructed, experienced and intersected by other identities in everyday life. By documenting everyday experiences in India it shows how religious identity does hold important implications for practices of citizenship and realizing everyday peace. This account is sympathetic to the studies by Peter Gottschalk (2000), Jackie Assayag (2004) and Katinka Froystad (2005) that problematize the notion that India's Hindu and Muslim communities are fixed and bounded entities by documenting the highly complex and composite experiences of religious identities, and the negotiated, fluid nature of encounters with "the Other." In order to think about the implications of religious identities in public life, the book illustrates how the politics of religion are spatially produced and reproduced through key sites, buildings, bodies and practices, and how differences and tensions are actively negotiated as well as connections and collaborations forged.

Crucially underpinning this line of enquiry is a concern not only with the everyday politics of religion but also the lived realities of Indian secularism. This account seeks to understand how the policy and ethics of Indian secularism are actually interpreted and mobilized, and how these practices shape both the aspirations for and realization of citizenship and everyday peace. Privileging a situated understanding of religion and secularism chimes with recent geographical research on religion that has called for "grounded theologies" that map the actually existing and complex realities of secularism (Tse 2013). It also expands our understanding of lived secularisms beyond prevalent European constructs (Iqtidar and Lehman 2012), which denote the privatization of religion (see Levey 2009) and stand in marked contrast to the Indian secular **imagination**, where religion is a fundamental part of the public sphere.

Structure of the Book and Key Arguments

As a body of research this study does not intend to offer prescriptions or predictions for how or when peace might be maintained. Rather, it seeks to uncover the ways in which peace was talked about, imagined and experienced by a section of the city's Muslim population who lived and worked alongside Hindu communities. Peace is an important lens for analysis precisely because Varanasi was popularly portrayed as a comparatively peaceful city. Had life in Varanasi in the mid-2000s been dominated by a rhetoric and reality of Hindu–Muslim violence, the starting point for interpreting Muslim lives and intercommunity relations would have been quite different. It is, therefore, vital that the argument of this book is located within its specific geography of peace.

The book explores everyday peace and citizenship through five empirical sites in the city. The backdrop to this research is outlined in the next chapter, which introduces the scalar politics of peace in north India by documenting experiences of violence, politics and peace within the state, city and neighborhood. The chapter includes a discussion about the methodological strategies employed and their implications for the production of the research. Chapter Three plays an important role in making peace visible in Varanasi by opening up the initial line of sight to different sites, scales and actors involved in the reproduction of peace in the city. It provokes some of the big questions of the book: how is peace situated, constructed and reproduced? And, in whose image is peace maintained? It does so by examining why everyday peace persisted in the aftermath of terrorist attacks on Varanasi in 2006, which were widely interpreted to be acts of violent provocation. The chapter documents the various actions of local, religious and political agencies which expose an important relationship between agency, "peace talk" and legitimacy when it comes to restoring everyday peace. It shows how peace is a process that is always being worked out through interactions within society and the state, and is complexly intertwined with the memory and anticipation of violence. And, it begins to illustrate how agency and narrative play out in a place which is heavily influenced by the shared intercommunity spaces of the silk industry and the uneven geographies of power that this embodies.

Chapter Four takes up this concern with the uneven geographies of power in Varanasi and turns attention to a Muslim-majority neighborhood to examine everyday encounters with the state and society, and the important role that policy and practice play in shaping everyday peace on the ground, from the margins. The chapter makes the argument that the reproduction of everyday peace is intimately related to the negotiation and articulation of citizenship for Varanasi's Muslims, where realizing citizenship directly concerns encountering and engaging with the state and society. Citizenship, therefore, provides a framework for thinking through the scalar politics of peace. In a nation where Hindu majoritarianism comprises the normative framework for public space I examine how Muslims strategically located themselves in both visible and less visible ways within public life, as they negotiated the possibility of realizing degrees of citizenship and justice. Chapter Five focuses on the civic spaces of Madanpura to show how local, national and geopolitical events have contributed to the sharpening of boundaries along religious lines whilst not entirely undermining practices of everyday coexistence. Within this context the chapter looks at the experiences and perceptions of inter-community interactions, and develops a perspective on how "playing with peace" can derive political and social capital for some. I document how Bengali Hindu processions annually interrupted everyday peace in Madanpura to momentarily destabilize peaceful relations and thereby heighten geographies of separation and interaction between the respective communities and police administration. By examining the formal and informal mechanisms deployed by state and society actors towards peace the chapter highlights contrasting experiences of security and insecurity and the uneven geography of peace as process.

In light of the city's uneven geographies of citizenship and experiences of peace and security, Chapter Six examines the spaces of economic peace in greater detail. By focusing on the experiences of Muslim Ansaris the chapter develops the argument that uneven geographies of power not only inform everyday peace, but are also critical for its reproduction. The chapter uncovers the ways in which imagined and real notions of "Hindu–Muslim *bhaīachārā*" (brotherhood) were experienced and perceived, especially by Muslim participants, and interprets everyday intercommunity peace as an ongoing process, which is simultaneously the by-product of economic conditions and ambitions *and* essential for the continued success of the market economy. The chapter draws central attention to the politics of peace and the role of the political within peaceful spaces, where the articulation of difference is continuously negotiated.

The book's earlier chapters argue that realizing citizenship and everyday peace often entails a struggle for inclusion, which is contingent on reproducing the status quo and unequal conditions of citizenship. Chapter Seven offers a contrasting view on citizenship in practice, which shows the limits of everyday peace, perpetuated through inequality and injustice. It mobilizes a particular event – a public protest against the arrest of a local religious teacher and concurrent framing, of both the Islamic scholar and the neighborhood, as "dangerous terrorists" – to show what happens when local Muslim residents chose to become visible and publicly defy the unjust actions of the state. I contend that the protest constitutes a particular act of citizenship that seeks to ask questions about justice, where local Muslims visibly constituted themselves as subjects of rights and contested the quality of citizenship. The appeals to reclaim citizenship and recast dominant stereotypes were articulated in ways that sought to reproduce, rather than jeopardize, the contours of everyday peace. The events described in this final empirical chapter serve to highlight a theme that runs through the book concerning the intimate and complex relationship between violence and everyday peace. And, it further questions the extent to which Muslim politics can be transformative within this particular geography of peace.

The concluding chapter returns to question what everyday peace means in this particular place. It draws out the key contributions of the book's arguments to emphasize the connections between space, place and agency and the contingent and precarious realities of remaking peace every day. It does so by reflecting on the empirical analysis in light of the broader postcolonial and geopolitical context of Varanasi and India to understand this particular geography of peace and its potential value for exploring peace as it happens elsewhere.

Chapter Two
The Scalar Politics of Peace in India

Interpreting peace in the everyday promotes thinking about the interlocking and multidimensional processes through which peace acquires meaning, as it becomes attached and reattached to certain sites, narratives and bodies and articulated through different scales. And, it describes the diverse infrastructures, processes and pathways of peace that shape society and space. In the Introduction I argued that the concept of citizenship provides an important framework for thinking about the scalar politics of peace, about the vertical relations with state and geopolitical practices and imaginations, as well as horizontal relations within society. With its focus on Muslim lives and lived experiences of peace, this geographical approach therefore concerns the interlinked matters of material and social inequality and living together across difference. The value of a geographical approach is that it takes culture *and* economics, institutionalized social difference *and* livelihoods into account (see Radcliffe 2007; Jeffrey 2007; Jones 2008). In her recent work looking at integration and exclusion in two French cities the geographer Kathryn Mitchell (2011) usefully emphasizes the mutual constitution of society and space and the interlinked nature of economic networking, urban development and cultural policymaking when it comes to maintaining peaceful urban life.

Everyday Peace?: Politics, Citizenship and Muslim Lives in India, First Edition. Philippa Williams.
© 2015 John Wiley & Sons, Ltd. Published 2015 by John Wiley & Sons, Ltd.

Therefore this study extends a perspective on peace that recognizes the importance of interacting agency and narratives across sites and scale in the (re)production of everyday peace.

In this chapter I illuminate aspects of the political, social, cultural and economic landscapes that were both the contexts and producers of forms of peace and uneven citizenship in north India, where research for this book was conducted, and through which insights have been interpreted. In order to ground Indian Muslim experience, the chapter is divided into sections that move between key interlocking sites: the nation, the state, the city, the market and the neighborhood. In each section, I seek to highlight the position of India's Muslims and elucidate something of the tensions, moments of violence and inherent inequalities that shape how practices and talk of "peace" find situated expression. In the final section I discuss how the methodological approach to fieldwork and matters of positionality informed the making of this research.

The Nation

India's postcolonial experience underpins its importance as a site through which to examine articulations of peace and violence. The divisive and violent religious politics that were exacerbated under British colonial rule and found brutal expression during the partition of the Indian subcontinent continue to inform the memory of, and potential for conflict between Hindu and Muslim communities. When, in the years immediately after India's independence, the future Constitution was debated in the Constituent Assembly, it was recognized that political structures were required to facilitate peaceful relations between the majority Hindu community and minority communities such as Muslims. This sensitivity around the potential for equality and peaceful relations was encapsulated by the constitutional commitment to secularism. Indian secularism provided a framework for managing differences between the Hindu majority population and religious minorities, and protecting their cultural rights and identities. In order to realize everyday peaceful relations, Indian secularism sought to ensure equal citizenship for religious minorities and to nurture an ethic of tolerance, connection and neighborliness between religions in a multicultural setting. As such, Indian secularism may be read as a road map for nurturing peaceful encounters between communities of difference, and as an aspirational document that encompasses the potential for equality and justice. But, in practice India's secular framework has been compromised and challenged.

The breakdown of peaceful relations during the Ayodhya affair in the early 1990s, and the Gujarat riots in 2002, fueled an extensive public debate about the evident failure or inadequacy of Indian secularism to successfully maintain "communal harmony" (Chatterjee 1997; Bhargava 1998; Hasan 2004). Some argue that the construction of secularism and the idea of the Indian citizen were limited from the outset: the secular state operating on the assumption that the Hindu majority was more secular and more equipped for citizenship than its minorities (Subramanian 2003, p. 137). In so doing, the Hindu majority re-inscribed itself as the universal, dominant group while assigning difference to minorities, that is, to subordinate groups. Although subordinate within the majority Hindu construction of India, India's Muslims, paradoxically, have been central actors in the project of nation building and (re)defining India's imagined and physical borders with the neighboring Islamic state of Pakistan. In many ways, the simultaneously ambiguous and contested status of India's Muslims undergirds the polarization of narratives around their membership in the contemporary Indian nation.

The reality is that despite, or even because of, secularism, India's Muslims are unequal citizens. Conflicting discourses at the national level prominently construct Muslims on the one hand as the "uncivil terrorist" and on the other hand as the "helpless victim." Such dominant rhetoric essentializes the figure of the Indian Muslim in very visible ways, whilst simultaneously hiding or denying the possibility of the Indian Muslim citizen, and casting aspersions on Muslims' loyalty to secularism and everyday peace. However, whilst secularism has suffered numerous critiques and has failed to protect against violence between majority and minority communities at the national level, there are signs that within everyday life, secularism continues to provide a powerful tool for imagining peace and coexistence, as I explore later in the book. I examine these realities more closely with a focus next on the state of Uttar Pradesh, which is home to the city of Varanasi.

The State

In this section I turn to the regional setting and the state of Uttar Pradesh (UP) in which everyday peace is situated. Political power in India is organized around a federal system of states which represent not only distinctive political characters, but also contrasting development and cultural realities (e.g. Singh and Srinivasan 2005). Uttar Pradesh is often cited as one of India's most important regions with respect to its demographics and role in national politics. Related to this it has a complex relationship with forms of

Hindu–Muslim violence, coexistence and development. I seek to elucidate some of that context here and show how it shapes the particular politics of everyday peace in north India.

Uttar Pradesh (UP) is India's most populous state (see Figure 2.1) and home to almost one-sixth of India's population (Government of India 2001). In part a corollary of its demographics, UP has traditionally proven to be the most crucial region in determining the formation of the central government in New Delhi; it sends 80 members to the Lok Sabha (the lower chamber of India's parliament) out of a total of 545. Until the end of the 1980s no party was able to form a union government without winning a majority of the Lok Sabha seats in UP (Raghuraman 2004). However, the national political significance of UP can also be attributed to the state's cultural politics, in particular, the major role that the region has played in the religious landscape of Hinduism and as the core of north India's "Hindi heartland." UP's towns and cities such as Mathura, Vrindhavan, Varanasi, Allahabad and Ayodhya are represented in the Hindu texts of the *Mahābhārata* and *Ramāyanai* (see Eck 1983), and the River Ganges, the sacred river of Hinduism, forms the backbone of the state as it flows southeast through the Gangetic Plains (see Figure 2.2). The region also has a rich Islamic history, from the end of the twelfth century through to the eighteenth century, the Gangetic plains formed the primary site for Muslim expansion in India. After Partition many Muslims chose to remain in India: Uttar Pradesh today is home to one-quarter of India's Muslim inhabitants. In statistical terms, Muslims make up 18.2 percent of the state's population, compared to the all India average of 12.4 percent. UP therefore also holds great meaning for Muslims, as the site of important political and religious organizations as well as prestigious educational centers, such as Aligarh Muslim University and historic qasbah towns, mosques, shrines and *imāmbārās*. Places such as Lucknow, Rampur, Faizabad and Jaunpur were also home to great musicians, dancers, Urdu writers and poets, as well as the focus of many creative and intellectual activities (Hasan 2001).

The interaction of Islamic and Hindu influences over the centuries has been made manifest in a composite culture experienced in architecture, music and literature as well as syncretic religious sects which extend across the Purvanchal region, including eastern Uttar Pradesh, Bihar and northern Madhya Pradesh. This shared culture importantly underpins practices and narratives of Hindu–Muslim coexistence in the region. Everyday evidence of this shared culture includes the *Bhakti* tradition led by Kabir in the fifteenth century, the vernacular Hindu epic produced by Tulsi Das and the nineteenth century Hindi–Urdu writings of Premchand. The poetry of Kabir, in particular, is frequently evoked in contemporary life to reference and actively

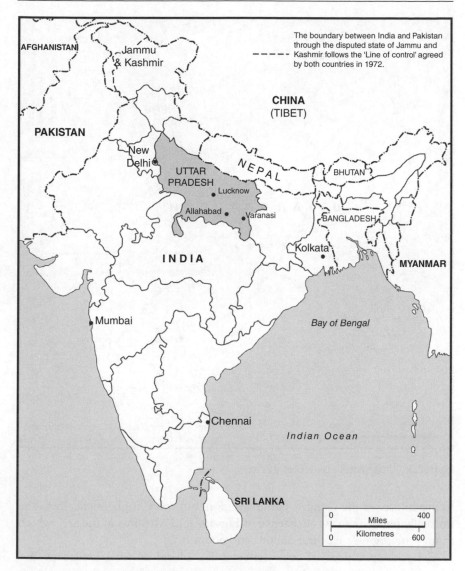

Figure 2.1   India, showing Uttar Pradesh and Varanasi. Source: Edward Oliver

reproduce everyday notions of intercommunity harmony within the region. Nongovernmental organizations, for instance, have drawn on the imaginary that surrounds Kabir in order to promote their agendas of intercommunity understanding and peace. This cultural co-production finds evocative expression in the term *Gangā Jamuni tahzēb/sanskritī*, which refers in a literal sense

Figure 2.2   Uttar Pradesh. Source: Edward Oliver

to the culture of the Ganges–Jamuna region, but more euphemistically to the mutually participatory coexistence of Hindus and Muslims in the area which shapes the region's cultural politics and economy.

Alongside practices of coexistence, UP's plural cultural landscape has become politically charged in recent decades, and this formed the context for so-called Hindu–Muslim violence. During the 1980s and 1990s, Hindu nationalist organizations such as the Bharatiya Janata Party (BJP) and members of the Sangh Parivar set about reworking the nationalist imagination in an explicitly Hindu image (see Kohli 1990, Manor 1997). Uttar Pradesh represented an important symbolic landscape in their project. A particularly important and contentious site was in Ayodhya where a mosque, the Babri Masjid, was built in the sixteenth century after the desecration of a

Hindu temple which Hindus believe was located at the exact birthplace of Lord Ram, one of India's most important Hindu Gods. The campaign to reconstruct a temple on the site came to represent the restoration of Hindu pride and spurred the growth of Hindu nationalism (see van der Veer 1994; Nandy 1995; Desai 2002). In September 1990 the BJP leader, L.K. Advani commenced a *rath yātrā* (chariot journey) across north India to Ayodhya in support of the movement. The *yātrā* provoked violence between Hindus and Muslims in towns across UP, though aspirations to demolish the mosque were not realized. Two years later, on 6 December 1992, the right-wing Hindu organization, RSS, organized a rally at the site of the Babri Masjid and invited BJP politicians and VHP members to give speeches. 300,000 members or sympathizers of fundamentalist Hindu organizations attended the rally which became increasingly uncivil and ultimately resulted in the demolition of the mosque. The repercussions of the event were widespread and on a scale not seen since Partition. The demolition triggered violence across Uttar Pradesh and north India more widely over the following four months and involved the deaths of approximately 1,700 people and the injury of at least 5,500 (van der Veer 1994).

These practices of religious violence underpinned by the cultural and political project of Hindutva rearranged UP's electorate along identity lines. After independence until the beginning of the 1990s the Congress party dominated in UP, with the exception of a very few, very brief periods. Its support reflected its inclusive caste and religious politics, incorporating Brahmins, Muslims and Scheduled Classes (SC). But the campaign to demolish Babri Masjid generated considerable sympathy for the BJP, particularly from the state's middle- and upper caste Hindus (see Subramanian 2003). In May 1991 the BJP won 34.5 percent of the vote in the state elections and a narrow majority (221 out of 425) in the UP Assembly (Wilkinson 2004, p. 161).

As the BJP mobilized the Hindu higher castes and some sections of the backward classes, Muslims grew disillusioned with the secular commitment of Congress and gradually shifted their loyalty to the Samajwadi Party (SP) under Mulayam Singh Yadav. Meanwhile SCs began asserting themselves under the Bahujan Samaj Party (BSP) led by Kanshi Ram and Mayawati (see Lerche 1999). Between 1991 and 2004 the UP electorate landscape was divided in three party political directions along class, caste and religious political fault lines which made coalition politics a common feature of UP state governments.

The Bahujan Samaj Party 's (BSP) landslide victory in the State Assembly Elections of 2007 marked a shift in state power, not only from the high castes to the low castes, embodied in the new Dalit Chief Minister, Kumari Mayawati, but also within the electorate. The party's inclusive caste formula brought Dalits, Muslims and Brahmins together using grass-roots strategies, blurring the support base across caste and religious boundaries, as well as between gender, education, generation and class (Hasan and Naidu 2007; Kumar 2007; Jeffrey and Doron 2012). The vote against Mulayam Singh Yadav's Samajwadi Party indicated popular resentment in the face of the incumbent government's failure to tackle the deteriorating law and order situation, as well as chronic development and infrastructure issues.

The 2007 Assembly Elections took place during my period of fieldwork and were held against a backdrop of increasingly divisive religious politics in UP, fuelled by the BJP's conspicuous communalizing tactics (Gupta 2007). Riot incidents in Mau in eastern UP in 2005 (see Rajalakshmi 2006) and in Gorakhpur and neighboring districts in 2007 were both allegedly linked to BJP member of parliament, Yogi Adityanath, and worked to create rifts between Hindus and Muslims in these areas. During the election campaign the BJP was said to have posted billboards in Muslim-dominated Aligarh with messages such as *"Kya inka irada pak hai?"* ["Is their intention pure?"] which played on the pun "Pak," meaning Pakistani. In this setting Muslims also voted decisively against the BJP in the 2009 Lok Sabhā elections. The legacy of events at Ayodhya continues to shape state political imaginations, contemporary politicians continue to make efforts to draw political mileage from the so-called "Mandir" issue, which is remembered annually in political and public spheres. Following the High Court judgment made on 30 September 2010 that the land should be divided into three equal parts (to the Infant Lord Ram, represented by the Hindu Maha Sabha; to the Sunni Waqf board; and to the Hindu religious organization, the Nirmohi Akhara) the dispute continues (see Patel 2010; Roy 2010; Menon 2011; Rai 2015).

Where UP's politics have been recently characterized by violence and progressive fracturing of identities, its development record is similarly bleak (see Jeffery et al. 2014). Since the 1990s UP's economy has been characterized by persistent economic stagnation and a deepening fiscal crisis, each compounding the other. This puts the state's economic growth of 4.4 percent way behind Gujarat, Haryana, Delhi and Bihar and only just ahead of Madhya Pradesh. The World Bank (2006) estimated that 31 percent of UP's population lived in poverty in 2000, much higher than the all-India average of 26 percent (*The Economist* December 11, 2008). Thanks to the fertile Gangetic Plains, UP is the largest producer of food grains (rice, wheat and pulses) in India, and agriculture comprises 78 percent of the workforce (Froystad 2005, p. 37). Other industries have traditionally included oil refining, aluminum smelting, production of edible oils, leather works, automobile tires, cement, chemicals, textiles and glass/bangles. Yet, even whilst these manufacturing sectors struggle, UP has failed to attract considerable foreign direct investment

(FDI) which, considering the scale of its territory and population compared to other states, is surprising. It has therefore played little part in attracting the newly emerging information communications technology (ICT), outsourcing and other "new industries," such as call centers, for which India has established a global reputation (Ahluwalia 2000 cited in Jeffrey et al. 2008).

The fiscal crisis in the mid 2000s led to a decline in public investment in the state, which in turn lowered the growth rate of the state economy (see Kumar Singh 2007). This economic condition resulted in a deteriorating public sector, a climate not conducive to supporting private investment and growth and a poor performance in the delivery of social and infrastructure services essential for growth and poverty reduction (World Bank 2002). The latter is reflected in UP's literacy figures which are below the national levels; in 2001, 70 percent of males and 43 percent of females over the age of 7 were literate, compared to all-India figures of 76 percent and 54 percent. Further reinforcing the connection between the decline in public provisions and the state's performance, a report on economic growth and its relationship with human development in UP identified "failures of public policy" as the under-lying reason for its poverty and backwardness. The key areas singled out include: poor quality of public services and inequity in access to them; visible deterioration in the functioning of public institutions; and understaffed and ill-equipped schools and hospitals, leading to decay in the public primary education system and health care in the state (Singh et al. 2005, pp. 10–11 cited in Pai 2007a, p. xxv). While a more recent World Bank (2006) report points to improvements in the state's fiscal health because of the imposition of financial discipline, these improvements have yet to be felt in the economy and polity (Pai 2007). Moreover different socio-religious communities in Uttar Pradesh experienced differential availability of, and access to, these welfare provisions. The mean years of schooling acts as one kind of development indicator where Muslims in particular demonstrated an espe-cially low result at the national scale and within Uttar Pradesh in particular. In a state noted for its development failures these have evidently impacted some communities, particularly the Muslim population, more than others.

Unequal Citizens?

As I have shown, the failure by successive governments in Uttar Pradesh to address issues of poor educational provision and under- and unemployment is intimately related to caste, class and religious inequalities in UP (Jeffrey et al. 2008). Although formally citizens of India, India's Muslims have not experienced the full spectrum of opportunities and rights that citizenship should, in theory, entail. As Hasan (2004) has argued, whilst Muslim cultural identity is protected under secularism, this does not extend to the safeguarding of Muslim economic and social rights, which have fallen woefully behind other socio-religious groups in post-Independence India, and contributed to their unequal citizenship (see Prime Minister's High Level Committee 2006). Literacy levels for Muslims stand at 47.8 percent compared to a state average of 56.3 percent, barely one percentage point above Scheduled Castes and Scheduled Tribes (SCs/STs) (Prime Minister's High Level Committee 2006, p. 287). On average, urban Muslim children attend school for a year less than all other populations, and for half a year less than SCs/STs. This reflects the failure of the UP Government to provide cheap accessible state education for Muslims in combination with the religious communalization of formal curricula, which has discouraged this community from investing in education in mainstream schools (Jeffrey et al. 2008, p. 44).

The fiscal crisis of the UP government, in conjunction with liberal economic reforms introduced in the early 1990s eroded government school provision, infrastructure and teachers, and contributed to the rapid growth of non-state educational facilities (see Kingdon and Muzammil 2001a, 2001b). This privatization of schooling also coincided with its politicization, particularly between 1999 and 2004. At the peak of the BJP-led coalition's dominance, the Hindu right worked to shape the educational institutions (Lall 2005) and curriculum in a Hindu national image, which necessarily excluded Muslims and Dalits (Jaffrelot 2007; Jeffrey et al. 2008).

In UP more widely, lacking the provision of government schools, poor Muslim communities are left with few affordable alternative educational opportunities to their local *madrasās*. A significant number of *madrasās* were established and funded by the Muslim elite during the colonial era as a means of protecting religious identities and fulfilling their religious duty to protect the poor. However, the growing dependence on *madrasā* education in contemporary India is believed to have done more to inhibit than to transform Muslim educational opportunities (Ara 2004). The potential for Muslims to act within circles of influence in UP has also been hindered by the decline in the use of Urdu in public spaces. This has left the language synonymous with Islamic religious education and marginalized in the north Indian public sphere (Abdullah 2002). The politicization of Hindi and Urdu in north India might have been driven by the elites (Brass 2004), but the negative impacts of Hindi's ultimate dominance in public spheres has been experienced by the poorest sections of UP's Muslim community (Prime Minister's High Level Committee 2006).

The Sachar Report also documented state–regional variations in Muslim conditions which showed that across most socio-economic and educational categories in UP, Muslims were on a par with, or below SCs/STs and "Other

Backward Classes" (OBCs). Meanwhile, intermediate Hindu castes and upper-caste Hindu elites occupied the top positions. Whilst SCs/STs and OBCs have been able to avail themselves of some of the benefits of education and employment reservations granted by the implementation of the Mandal Commission in 1991 (see Jaffrelot 2003), comparative reservations were not extended to marginalized Muslims. Moreover the notion of "Muslim Reservations" has not captured the imaginations of the Muslim masses. On the one hand a group of backward-class Muslims already enjoys the benefits of the Reservation policy of state and central governments, who are not inclined to extend this privilege to all Muslims. On the other hand, there is a real fear that reservations for the Muslim community at large would produce a vehement backlash from India's Hindu community, comparable to that witnessed in the aftermath of the Shah Bano judgment (Alam 2003, p. 4881). Wary that reservations constitute a double-edged sword for Muslims, a recent Minister for Minority Affairs, Salman Kurshid, spoke progressively about alternative angles for redressing Muslim inequality such as the potential of the private sector for inclusion (*The Hindu* June 21, 2009). Incremental improvements in employment and educational arenas were bolstered for Dalits by their increase in political power through the expanding capillaries of the BSP. This was gradually reinforced through the party's brief spells of control in the state between 1993 and 2003, before momentously coming to power in 2007. The BSP has attempted to place Dalits in key positions within the UP bureaucracy and improve their access to police protection and judicial redress (Jeffrey et al. 2008, p. 42). But UP's Muslims have experienced little assistance in state administrative and political arenas under Mulayam Singh Yadav's terms of office, despite the Samajwadi Party's 2004 Assembly Election promises (Engineer 2004).

Despite being disproportionately concentrated in urban areas, since independence, India's Muslims have been significantly underrepresented in India's capitalist elite (Harriss-White 2005, p. 173), as corporate executives (Goyal 1990, pp. 535–44 cited in Harriss-White 2005, p. 173) and in the various institutions of the Indian state, such as the central government, the railways and the armed forces (Khalidi 2006, pp. 41–7; Prime Minister's High Level Committee 2006). Like India's other marginalized communities, India's Muslims are overwhelmingly represented in occupations in the "informal economy" (Prime Minister's High Level Committee 2006), the sector operating in contrast to, but not independently of the "formal economy" but typically out of the direct reach of the state (see Harriss-White 2003; de Neve 2005). In urban areas Muslims are more likely to be self-employed than Hindus (Prime Minister's High Level Committee 2006, p. 91), often preferring to be self-sufficient and running small, low-investment businesses which, in general, are outside, if not wholly excluded from, the new economy in India (Hansen 2007, p. 51).

Craft industries are one area where Muslims have traditionally played a distinctive role (Harriss-White 2003; Khalidi 2006; Prime Minister's High Level Committee 2006). In Uttar Pradesh particular cities constitute distinctive centers of "Muslim professions"; in Moradabad Muslim artisans produce brassware (see Hasan 1991; Ruthven 2008), in Ferozabad glassware, in Khurja pottery, while carpets are manufactured in Bhadodi and Mirzapur, hand-printed textiles in Farrukhabad, *chikan* embroidered clothes in Lucknow (see Wilkinson-Weber 1999), cotton and silk embroidery in Varanasi, and handloom cloth in Mau (Khalidi 2006, p. 89).

Within the national historical context the figure of the urban Muslim artisan has been materially and imaginarily marginalized. The *swadeshi* movement in the decades leading up to independence constructed the compelling image of the rural, typically Hindu, artisan at the center of India's independence struggle, and the primary beneficiary of subsequent post-colonial rural development (Mohsini 2010, p. 195). Meanwhile, the discursive production of "craft community" and "traditional Indian craft" employed by government and development agencies as well as consumers, denies agency to artisans regarded as subjects of development interventions in the case of the former, and craftsmen from a timeless era in the case of the romantically inclined latter (Venkatesan 2009).

Narrow bands of wealthy Muslims do exist in UP's towns and cities, represented by the princely retainers of the Awadh/Mughal kingdoms, old trading elites, or where some have demonstrated considerable economic success through entrepreneurial and capitalist enterprises. In Aligarh, Muslims own iron-lockmaking industries and have also ventured into producing building materials (see Mann 1992) and in Varanasi some sections of the Muslim weaving class established a hold over the silk sari trade (Hasan 1988, p. 833; Engineer 1995). Muslim capitalists in Moradabad have reoriented their brassware industry to satisfy global markets, most notably exporting to rich Arab countries (Khalidi 2006, p. 89; Ruthven 2008).

The urban economy therefore represents one arena in which India's Muslims have traditionally encountered and interacted with people from all backgrounds and walks of life. As a locus for intercommunity interaction, the economy presents the possibility both for cooperation (Sennett 2011) *and* conflict (Chua 2003), and can influence the nature of material and imagined relations beyond the immediate workplace. Of course, the broader institutional and political environment informs the nature of local economic dynamics. In India, Harriss-White (2005) has expressed anxieties about the way in which the Indian state's secular aspirations have left the economy open to the influence of religious communities and identities, which may often lead to competition and the entrenchment of hierarchies. Yet, little research has sought to illuminate how big politics interacts with local economic practices, and ultimately what this means for everyday practices of citizenship. The ways in which religious identities inform and shape economic transactions, as well as moments of encounter and realizations of citizenship within a particular place, are the focus of Chapter Six.

The City

As places which "pose in particular form the question of our living together" (Massey 2005, p. 150), cities constitute important sites through which identities are staked, ideas about difference and belonging are negotiated and rights are claimed. Such diversity means that cities are often marked by processes of exclusion, segregation and repression (Peach et al. 1981; Holston 1999; Mitchell 2003) as well as real and imagined violence (Gregory 2006; Graham 2010). But, cities and their urban publics have also been celebrated as arenas for nurturing tolerance, coexistence and everyday peaceful realities; as places in and through which cosmopolitan and multicultural practices find expression, at different periods through time (Hanf 1993, Milton 2009).

There has traditionally been a rich appetite for understanding social and political life in the context of Indian cities, made manifest in research on urban localities as sites of public culture (Freitag 1989, 1992; Haynes 1991), labor relations (Chandravarkar 1994), social change (Searle-Chatterjee 1981), political challenge and control (Hansen 1999), and state control over the poor (Gooptu 2001). More recently, since the "urban turn" (Prakash 2002), cities have been reorientated towards the center of analysis, particularly in research on the production of violence (Hansen 2001), the emergence of India's new middle classes (Varma 1998; Fernandes 2000, 2008) and the interaction of neoliberalism and space, for instance in the narrative construction of slums (Roy 2002; Ghertner 2011) and the "misrecognition" of the poor (Bhan 2009). This work has typically examined metropolitan centers (Mumbai and New Delhi) and the impact of global communication, information technologies and economic restructuring on urban life. But, it is not just the mega-cities, those otherwise on the map (Robinson 2002), that are proving important sites for understanding the intersection between urban processes, social relations and change; India's regional towns and cities are integral too (see Olsen 2003, 2005; de Neve 2006; Jeffrey 2007; Ruthven 2008). Particularly if we want to understand politics of everyday peace, then focusing on ordinary life in "ordinary cities" (Robinson 2002) is paramount.

Yet, "ordinary cities" also constitute sites for the extraordinary, and Varanasi is certainly exceptional in some ways. Often imagined and represented as India's most sacred Hindu pilgrimage center (see Eck 1983), concurrently, its social and cultural urban spaces have been commonly examined through the imagined and lived realities of Hinduism (Hertel and Humes 1993; Parry 1994, Singh and Rana 2002). But Varanasi is also home to a sizeable Muslim population, which in 2001 comprised 30 percent of the city's residents, significantly more than the percentage of Muslims in UP (Government of India 2001). Unlike the city's majority Hindu inhabitants (63 percent), who hold a range of occupations in different economic sectors, Muslims in the city, predominantly Sunni *Ansāris*, are involved in the production of silk fabrics, as well as other smaller artisanal industries (see Kumar 1988). In India's caste based hierarchy, occupation and work are intimately tied to social status. This is particularly the case for lower-caste/class groups, as has been documented in Varanasi with respect to sweepers (Searle-Chatterjee 1981), artisans (Kumar 1988), funeral attendants (Parry 1994), boatmen (*mallahs*) (Doron 2007, 2008) and washermen (*dhobhīs*) (Schutte 2006).

Muslims first settled in Varanasi in the eleventh century, when, following the defeat of an invading Muslim army, women, children and civilians were permitted to remain on the northern side of the city and serve the Hindu kings. Many learned the craft of weaving, incorporating their skills and designs into the fabrics. Jean-Baptiste Tavernier, the French explorer and cultural anthropologist, visited Varanasi between 1660 and 1665 and reported that in the courtyard of a rest-house in the Chowk area the trading of *reshmi* (silk) and *suti* (cotton) fabrics was taking place between Muslim *kārighars* (artisans or craftsmen) and Hindu *Mahajāns* (traders) (Chandra 1985, p. 177). During the post-Mughal and colonial period, when Varanasi underwent a period of Hindu revival (Dalmia 1997), power was largely held by the nexus of the local Rajput dynasty, the merchant-bankers and the "*gosāīns*" ("mendicant trader-soldiers"). However space was also created for lower class Muslim weavers in the city, who were dependent on close relations with Hindu traders (Bayly 1983, pp. 181–4; Freitag 1992) on the one hand, and the subject of Islamic reformist activities on the other (see Gooptu 2001, pp. 305–12).

Compared with some cities in UP, such as Meerut, Moradabad, Aligarh, Allahabad and Ayodhya, Varanasi has witnessed comparatively few incidents of Hindu–Muslim violence, and when it has witnessed riots they have been relatively low intensity in terms of fatalities (see Parry 1994, p. 35).[1] Since Varanasi's most serious riots in 1991 and 1992 (Malik 1996), there have been no large-scale incidents of violence between Hindus and Muslims, but the city, like many in UP, has not been free from moments of interreligious tension. Varanasi is a center of politicized "neo-Hinduism." The BJP is dominant in the local Banaras Hindu University (BHU) and along with the Rastriya Swayamsevak Sangh (RSS), has a notable power base in the city (Searle-Chatterjee 1994b; Casolari 2002).[2] Hindu nationalists have attempted to draw upon the religious symbolism of Varanasi to mobilize its supporters around allegations that the Gyanvapi mosque was built on the grounds of the Vishvanath Mandir (temple) after its destruction by the Mughal emperor, Aurangzeb in the seventeenth century. Given the events at Ayodhya in the early 1990s, their threats to desecrate the mosque are taken seriously by the city's administration, which has deployed police forces to guard the complex at an annual cost of Rs18 crore (10 million) for the past sixteen years (*Times of India* February 11, 2009).

The dominance of local Hindu right personalities and politicians was apparent in 2002 when they forcefully intervened to halt the filming of "Water" directed by Deepa Mehta. The film plot describes the plight of Hindu widows condemned to poverty at a temple in Varanasi in the 1930s, as such a critique of social restrictions and oppression suffered by India's Hindu widows. Although the script was approved by a national censor board, individuals and groups closely associated with the RSS claimed that the film besmirched India and constituted part of an organized plot by the Christian church against Hinduism. As filming got under way at Assi *ghāt*, a mob of Hindu fundamentalists tore down the set. The violent objections caused the UP government to retract its permission to film, fearing the escalation of violence in the city (see Saltzman 2006).

Alongside moments of tension and violence, narratives of harmony and cooperation have continued to shape the city's public sphere and image. For instance in March 2005 a devout Muslim woman, following orders from the God Lord Shiva, constructed a temple in the Gangeshpur locality. She received considerable moral and financial backing towards her project, from both the Hindu and Muslim communities who supported her wish. Muslims participate in the Hindu Ramlila celebrations, taking the parts of Hindu Gods as well as being involved in set production. Figures such as Bismillah Khan and Nazir Banarasi were publicly projected as engendering *Gangā Jamuni tahzēb/sanskritī* and thereby uniting both communities through their music (Hunt 2006), and poetry respectively. Such narratives of interreligious harmony circulate widely, and are reproduced within national as well as local imaginations. The city's silk sari market is often cited as a key explanation for, and evidence of, these good relations because of its involvement of both Hindus and Muslims, and it is this site that I turn to next.

The Market

Forms of cultural co-production and notions of harmony between Hindus and Muslims interlink with economic practice in the city, most notably in the spaces of the silk sari market. Varanasi is one of UP's most famous handloom centers, especially for silk saris, and in particular the Banarasi brocade (see Agrawal 2004) (see Figure 2.3). In addition, it is known for the production of carpets, wooden toys, *Banarsi Pān* (betel leaves) and *Banarasi Khoyā* (milk product). In 2001, industry and manufacturing employed just 11 percent of the population, whilst 51 percent of workers were engaged in household industries such as weaving and spinning. The other important sectors are metal and manufacturing (15 percent), printing and publishing (6 percent) and electricity-machinery (5 percent) (Kumar Singh 2007). With the exception of some small-scale call centers on the outskirts of Varanasi, the city has largely failed to attract the economic benefits of information communication technology or high-tech industry associated with "modern" India. Consequently, ambitious young students from Varanasi tend to search for better-paid and more dynamic jobs outside of the city, in places such as New Delhi, Hyderabad, Bangalore, and Mumbai.

Figure 2.3   Banarasi Brocade. Source: Author (2006)

Like many cities in north India in recent decades, Varanasi's consumer landscape increasingly bears the mark of private investment. The introduction of "Western" names catering to new tastes sometimes jarred with the sentimentalities of long-standing Banarasis, who were unhappy to the see the city being transformed into a Western image. Despite some opposition to the hugely hyped opening of a McDonald's in a brand-new shopping mall, for the growing number of middle-class families this was perceived as an exciting leisure alternative. Once heralded as India's "most picturesque city" (Caine 1890 cited in Huberman 2010, p. 176) domestic and international tourism is big business in Varanasi, creating opportunities for savvy entrepreneurs, such as the boatmen operating from the riverfront (Doron 2008) as well as challenges for others caught up in the tourist economy (Huberman 2010). Numerous guesthouses have colonized old *havelīs* (private mansions) along the *ghāts* where shops and restaurants now cater to foreign tastes. Meanwhile, the development of new luxury hotels in Cantonment cater for the influx of tourists now able to access the city on a one-hour flight from New Delhi.

The silk sari industry is based on a network of small to medium scale *kārkhānās* or workshops, concerned primarily with weaving, dying, embroidering and polishing; resembling a "decentralized factory" (Chari 2004,

Figure 2.4   A weaver carries silk *belans* whilst others dry in the sun. Source: Author (2007)

p. 760) (see Figure 2.4). These networks feed into extensive trading markets that link the city across state, national and international spheres. Three broad categories of enterprise may be understood in terms of their function or organization: first, manufacture or production; second, commission agents or middlemen otherwise known as *girhasts*; and, third, dealers and traders locally known as *gaddidārs or kotīdārs*. The manufacture or production of silk fabrics in urban Varanasi is largely confined to some Muslim majority neighborhoods such as Alaipura, Lallupura, Pillicoti and Jaitpura in the north, Madanpura and Kalispura in central Varanasi and Bazadhia on the southwest side (see Figure 2.5).[3] Males, typically aged 15 to 45 years but sometimes older, carry out the majority of weaving and dying work, often with the assistance of younger boys and girls, whose dexterity is valued in weaving the most intricate designs. The *kārkhānās* are located on the ground floor of four- and five-storey houses which line the narrow *galīs*, and with the wooden shutters invariably flung open, the clack-clacketing noise of looms forms the background rhythm of these weaving neighborhoods. Female input in the industry is ubiquitous and extensive in scope within domestic spaces. From keeping business accounts to spinning yarn, preparing bobbins and cutting excess *zarī* (gold or silver thread) as well as household management and childcare, females of all ages are the silent backbone of the industry (Kumar 2007). However women remain

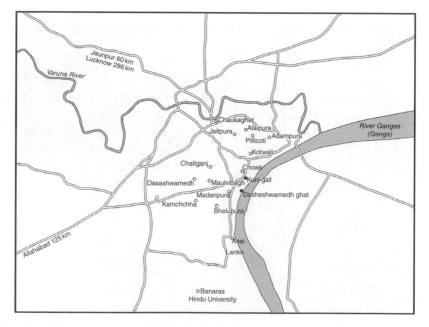

Figure 2.5   Varanasi, showing majority Muslim neighborhoods. Source: Edward Oliver

conspicuously invisible within the public spaces of interaction that constitute the marketplace and which represent the focus of research in Chapter Six.

Business transactions have been conventionally located in the market at Kunj gali, located in Chowk, central Varanasi. This is the city's oldest sari *mandī* (market), dating back over 400 years, and the largest, with an estimated 500 trademark-pink *gaddīs* (shops) lining a maze of narrow *galīs*. Here, shop owners or *gaddidārs* are primarily Hindus (mainly from Aggarwal, Jaiswal and Marwari families), as are the commission agents who mediate the transactions at multiple levels between producers, buyers and sellers. While Kunj gali attracts individual customers, often Hindu ladies visiting the city on pilgrimage, as tourists or even for work, like the Minister from Rajasthan, the market is predominantly a wholesale outlet catering for sari businessmen from across UP and India more widely. Opening Monday to Friday, the market operates its most intensive trade in the late afternoon and early evening when the lanes buzz with activity: men huddle in close discussion over the price of saris spread out before them, whilst porters and brokers move swiftly through the lanes with large bundles atop their heads or smaller packages tucked under their arms.

Away from the central hub of this sari market, smaller satellite markets have developed over the years in the traditional weaving *mohallās* (neighbor-hoods). Unlike at Kunj gali, traders in these satellite markets are mostly Muslim – Sunni *Ansāris*, descendents of traditional weaving families. In the majority Muslim *mohallās* of Madanpura, Lohta and Pillicoti trade also increasingly takes place with businesses across the Indian states and abroad. Since the 1970s, Varanasi's oldest Muslim weaving neighborhood, Madanpura, has established a reputation as an important market for the purchase and sale, as well as the production, of silk saris. Muslim *Ansāris* have therefore established extensive economic networks which extend beyond the city, state and national boundaries, as they sell to wholesale enterprises located in India's large metropolitan centers, and occasionally trade directly with buyers in the Middle East, America and Europe. (Given that Muslim *Ansāris* work in a range of different roles within the industry I use the term Muslim *Ansāris* when referring to the Muslim community at large, but otherwise employ occupational references to make distinctions clear.)

Between 2005 and 2008, there was an overwhelming perception amongst Muslims and Hindus, weavers and traders alike that the silk sari *handloom* industry was in a state of decline. Processes of trade liberalization and globalization within India (see Corbridge and Harriss 2000) had prompted significant restructuring of the industry. In particular, demand for Banarasi handloom goods was perceived to be down because of the heightened competition they faced due to the import of cheaper Chinese-produced silk yarn and imitation fabrics. The rising price of raw materials, particularly of handloom silk yarn reduced the capacity of handloom weavers to participate competitively in local and global markets. Meanwhile, imitation Banarasi saris produced on power looms in large domestic centers such as Surat, Gujarat easily undercut the handloom market on productivity and price. The Indian government has implemented a number of schemes to aid weavers over the years (see Roy 1993; Srivastava 2009), most recently the Integrated Handloom Cluster Development Scheme and the Health Insurance scheme, but these are often not fully realized by the weavers themselves. Low levels of weaver education and awareness, in combination with processes of marginalization by wealthier traders and cooperative owners, inform this situation. Consumption patterns are changing, too, as women shift away from weighty expensive Banarasi brocades to lighter and more manageable embroidered outfits. In urban centers women increasingly preferred the *shalvār kamīz* (loose fitting trousers worn with long shirt) or "Western" styles of dress (Guha 2004 and for a discussion about the importance of dress and identity in north India see Tarlo 1996) inspired by actresses in television serial dramas such as *Kahānī Ghar Ghar Kī* and *Kahin To Hogā*. But where patterns of globalization brought competition and decline in the handloom product, marking despair for weavers, some, with sufficient economic and social capital, have been able to diversify and continue to compete in the changing fabrics industry, responding to a demand for power loom, embroidered and printed fabrics as well as dress material and household furnishings.

The predicament of Muslim weavers in Varanasi has been touched upon by scholars (Kumar 1988; Showeb 1994; Ciotti 2010), commissioned reports (Srivastava 2009), national and international media (Mukerjee 2004; *The Economist* 2009) and novelists (Bismillah 1996). They collectively document the idea of mutual interdependence that structures relations between weavers and trader, Muslims and Hindus, and simultaneously highlight widespread practices of discrimination against Muslim weavers. More recently, Raman (2010) has studied the character of relations between Hindus and Muslims in the silk sari industry with a particular focus on gender. This book develops an insight into Hindu–Muslim relations in Varanasi in order to understand the experiences of Varanasi's Muslims with respect to the horizontal relations in society that frame practices of citizenship and the production of everyday peace within the spaces of the nation, city, market and neighborhood.

The Neighborhood

The "locality" is a productive setting through which to understand Muslim identity and experience. Representing the materialization of spatial concepts, memories and practices that shape social relationships, the locality therefore forms the site where individual agency becomes articulated (de Neve and Donner 2006, p. 10). As the locality that connects direct experience and knowledge of households and families with their participation in city, national and global networks, the neighborhood is a methodologically productive field site. The interpretative paradox to bear in mind is that neighborhoods are the producers of contexts, just as contexts produce neighborhoods (Appadurai 1996, p. 184). In seeking to understand Muslim experience in the city, I dedicated a significant proportion of research energy to spending time with and interviewing Muslim residents in one of Varanasi's major Muslim neighborhoods, Madanpura (see Figure 2.6). This book builds on the insights presented in a recent edited collection by Gayer and Jaffrelot (2012) that examined Muslims in Indian cities, but notably lacked a perspective from Varanasi.

Madanpura is a traditional weaving locality in the center of the city, home to approximately 40,000 Muslims. Like Varanasi's wider Muslim population, the vast majority of residents in Madanpura are Sunni *Ansāris*, the majority of whom are engaged to some degree in the silk sari industry. *Ansāri* means "weaver," but the title also represents a history of struggle to differentiate *Ansāri* identity from the *Julāhā* caste of handloom weavers, fought in the early twentieth century (Pandey 1983, 1990; Kumar 1988). Gooptu (2001)

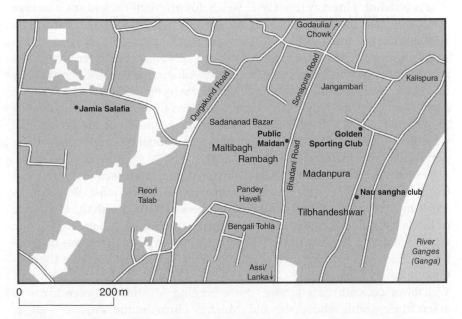

Figure 2.6   Madanpura, a Muslim majority neighborhood. Source: Edward Oliver

has shown how this shift in identity was coupled with forms of religious and occupational assertion for *Ansāris*. These assertions found public expression through Muslim celebrations and festivals such as Muharram,[4] as well as a diverse range of activities sponsored by *ulema* (Islamic scholars) and elite Muslim merchants that included the provision of religious teaching, *anjumans* (assembly, group) and *akharas* (wrestling and training arena), the protection of *waqf* (religious endowments) properties, and the provision of relief in Muslim areas during riots in the early twentieth century (Gooptu 2001, pp. 73–4). Such actions were provoked by feelings of deprivation and dispossession in the face of resurgent Hinduism in north Indian cities at that time, especially in Varanasi (see Dalmia 1997; Gooptu 2001). They were situated within a broader discourse concerned with the decay and decline of Indian Muslims, but also of Islam worldwide in the interwar period. Muslim weavers came to recognize their condition as relatively more deprived than the poor of other Indian communities (Gooptu 2001, p. 268).

Local residents suggested that to the north and east of the Bhadani road lived Madanpura's older, wealthier and more refined families, many of whom owned trading enterprises. Immediately to the west of the main street were relatively newer, but expanding, silk sari firms, and in the west towards Reori Talab lived some of the neighborhood's poorest weavers. In fact, the large majority of Madanpura's population occupy lower socio-economic groups. The population's literacy levels were well below the national and state average at 65.8 percent for the population, 57 percent male and 43 percent female (Government of India 2001). Residents of Madanpura only married within neighborhood families and regarded other wards as poorer, inferior and less meticulous in Islamic observance (see Searle-Chatterjee 1994b). The majority of weavers were Sunni, of whom a significant proportion belonged to the *barelwi* sect (approximately 40 percent), followed by *deobandi* (approximately 35 percent). Meanwhile the most visible group to those outside the *mohallā* were the *Ahl-i-hadiths*, who accounted for the smallest proportion of the population (approximately 15–20 percent).

Between the 1930s and 1950s, a cluster of merchant weavers converted from the *deobandi* sect to the more conservative *Ahl-i-hadith* (Wahabi). In 1966 the Arabic University, Jamia Salafia, was established with the assistance of funds from Saudi Arabia. This now forms the largest college for Islamic education in the city, with over 500 students. Many of the religious instructors at the school receive their religious education in Saudi Arabia (Searle-Chatterjee 1994b). Searle-Chatterjee (1994b) argues that the move to Wahabism constituted a strategy for emerging Muslim silk merchants to assert their status above the old Muslim elites in the city, as well as distinguish their merchant identity from poorer weavers. The relatively strict conservative approach to Islam embodied in the Ahl-i-Hadith Islamist reformist school dominates public impressions of Madanpura held by the majority Hindu community. Despite the sect's relatively small percentage of the neighborhood population, Madanpura, nonetheless, was widely regarded within Varanasi as a particularly conservative Islamic area.

During my fieldwork, the streetscapes of Madanpura's two main thorough-fares, along Durgakund Road and Bhadani Road, were characterized by an intensity of business activity, largely silk fabric shops and showrooms. Many were extensive; air conditioned with smokescreen windows to protect patrons and customers from the sun and curious onlookers. There were also dye merchants, parcel shipping agents, sweet shops and *chāy* (tea) shops. Above the shops, homes extended upwards, and up again as new levels were added to accommodate growing families in a densely developed *mohallā*. On the main high street (Bhadani Road), large *havelīs*, including the family home of the late poet Nazir Banarasi, reflected the history of wealth amongst a narrow band of the neighborhood's older trading families.

Local inhabitants, traders and "touts" or commission agents, businessmen, visitors, pilgrims and tourists, men, women and children representing all castes and religions flow through the neighborhood in rickshaws, cars, buses and jeeps, on bicycles and on foot. In addition to those visiting friends and relatives in the *mohallā*, others, involved in one of the many trades or aspects of the silk industry, and coming from beyond Varanasi, the state and outside India, may stop for business and disappear behind shop doors. Meanwhile commission agents looking for cheaper, perhaps defective or "seconds" goods (*lāt ka māl*) head into the narrow *galīs* leading off the main street.

Whilst Varanasi's history is shaped by relatively low intensity and infre-quent conflict between different sections of the city's Hindu and Muslim communities, Madanpura's residents have, at times, been implicated in these incidents. Moments of tension and conflict between Madanpura's residents, local Hindus and the administration offer a lens through which one may understand how more extraordinary events shape and are shaped by everyday peaceful realities. Festival processions by religious parties through another religious community's neighborhood have long constituted a source of contention in different parts of India, as they do in Madanpura. While Islamic Muharram processions have occasionally generated tense incidents in the city involving attacks by Hindus on Muslims (see PUCL 2000), the Durga and Kali Puja processions conducted by local Bengalis through Madanpura have created annual cause for concern since the late 1970s, when the Durga Puja ended in violent Hindu–Muslim conflict in 1977 and 1979 and alter-cations during the Kali Puja led to inter-community riots in 1991 (see Khan and Mittal 1984; Engineer 1995; PUCL 2000).

Such incidents were significantly shaped by the police administration, which sought to control the situation by implementing curfews to enforce spaces of separation between different communities. However the police were also accused on different occasions of acting with prejudice against Muslim *Ansāri* residents, thereby exacerbating rather than alleviating the impact of the conflict. The greatest threats to peace in contemporary Varanasi were understood to emanate from Muslim *Ansāri* neighborhoods. This perception resonated with historical reports and attitudes towards Muslim *Ansāris* in early-twentieth-century UP that constructed this group as a menace to society's peace, health and prosperity (Gooptu 2001, p. 261; see also Pandey 1983). These civic interactions, though episodic, also capture a particular way in which intercommunity relations are framed in Madanpura, and India more widely.

The matter of representation is important for understanding how notions of difference gain circulation within society. India's print media has played a significant role in constructing relative notions of trust and suspicion between Hindu and Muslim communities at different times. During the tension surrounding the Ayodhya issue in the 1990s, the north Indian Hindi media was widely perceived to be prejudiced against Muslims, and propagated falsehoods that wrongly implicated Muslims in acts of violence against Hindus, heightening local atmospheres of fear, insecurity and violence (Rawat 2003). In Varanasi, two local Hindi dailies in particular, *Aaj* and *Dainik Jagran*, became the mouthpieces of Hindu right organizations, as their proprietors and journalists were intimately linked to organizations on the ground (Malik 1996). Deepak Malik, a local scholar and activist, described *Aaj* as the "biggest machinery of disinformation" and recalled how its editor would encourage flagrant lies to be published. For instance, following the destruction of the Babri Masjid the paper published rumors that widely exaggerated the number of Hindu deaths and provoked violent reactions by Varanasi's Hindu youths. Had information been more responsibly disseminated Malik believes this situation would have been avoided. More recently, the tone of the media has become arguably less sensational and more sensitive around Hindu and Muslim matters, as Chapter Three suggests.

Field Research

This book draws principally on doctoral research conducted over thirteen months between 2006 and 2008 and on shorter, month-long, post-doctoral trips in 2010 and 2011. I was initially driven by an interest in understanding everyday Hindu–Muslim encounter and exchange in the city of Varanasi in north India. In particular I wanted to examine the spaces of the silk sari industry in the city, which was frequently heralded as the principal site of "communal harmony." Then, in March 2006, whilst I was learning Hindi in Varanasi, twin terrorist attacks were carried out in the city; one bomb was detonated in a popular Hindu temple and a second at the main railway station. Islamist activist groups were immediately suspected of being behind the bomb blasts and an atmosphere of anticipated Hindu–Muslim tension prevailed in the city, as many residents, politicians and the media feared the potential for a Hindu backlash or rising intercommunity tensions.

In the aftermath of the blasts I drew upon the research assistance of a local resident, Hemant Sarna, to conduct a small project that examined attitudes towards the blasts and why peace prevailed. We interviewed a spectrum of people from different socioeconomic and religious groups living largely in central and south Varanasi as well as political and religious leaders and solicited the opinions of journalists in tandem with local and national written material (see Chapter 3). That intercommunity violence did not result, and was actively averted in different ways, proved fascinating and prompted me to take a much broader perspective on everyday intercommunity life in the city. The incident worked to make visible Varanasi's Muslims and highlighted the disproportionate burden for maintaining the peace that appeared to fall on their shoulders. Intrigued by how Muslims were positioned within society, and orientated towards processes of peace and citizenship, the overall research question shifted to focus more closely on Muslim experience and agency in the city's multicultural urban life.

I chose to base a large part of my research in the Muslim majority neighborhood of Madanpura. Located in the center of Varanasi this neighborhood bordered majority Hindu localities and played a dominant role in Hindu imaginations about Muslims in the city. Madanpura had gained a reputation as an established sari market, but its conspicuous Muslim identity also represented a site of the "Other" within the city. Moreover, two main thoroughfares intersected the neighborhood along a north–south axis, bringing residents, businessmen and women, tourists and pilgrims to and through the neighborhood. Working in this context it might have seemed intuitive to turn to a local Muslim resident for research assistance given their familiarity with this area. In the event, whilst I interviewed local Muslim *Ansāri* residents, in particular a former journalist and social activist, no one demonstrated the kind of expertise, sensitivity and energy towards the project that Ajay Pandey (known as Pinku) displayed (see Figure 2.7). Indeed, it was his distance from the research site which made his contributions particularly sharp and valuable. Pinku was introduced to me by Rakesh, the proprietor of Harmony

Figure 2.7   My research assistant, Ajay Pandey. Source: Alexandra Daisy Ginsberg (2008)

bookshop, which served as a popular hub for Western scholars, writers and researchers in the city. Together, Pinku and I embarked on the project. Despite growing up on Assi ghat, less than three kilometers from Madanpura, Pinku was less than familiar with the Muslim majority neighborhood and even less familiar with what life as a Muslim and as a weaver was like in Varanasi. Pinku had Muslim friends from his school days, but as he had grown older, became busy in work and later married life, these ties had weakened. Such circumstances were replicated across the city, especially in Brahmin Hindu homes where relations with Muslim families were rarely intimate and often entirely absent.

While Pinku was unfailingly professional in the field, relentlessly amiable and always humble and courteous to our informants, there was still space for him to play on the stereotypes of Hindu and Muslim mistrust. In the early stages of fieldwork, on a day when we were still familiarizing ourselves with the dense network of *galīs* in the old part of the Muslim *mohallā*, Pinku joked about his sense of vulnerability, saying that I should sign a declaration so that if anything fatal happened to him while working with me, I would agreed to support his family. Of course he was just joking, but nonetheless this comment did reveal an ingrained sense of apprehension towards the Muslim community. It was not until after a few months working together that Pinku acknowledged the initial unease he had felt about our working environment.

As we fostered our informants' trust and gained an insight into their lives and views, Pinku often remarked at the absurdity of his original views and those of his immediate Hindu community, who lived in the same city alongside Muslim communities, but who knew so little about the reality of their lives.

My living arrangements also offered a fascinating window on to everyday life in the city. During the period that I lived and returned to Varanasi I stayed with a Brahmin Hindu family, the Tripathis, in their house in Lanka. My initial intention was to stay as a paying guest whilst learning Hindi before moving on to an apartment in the center of the city and closer to Madanpura. However, as my relationships with the family quickly developed it became impossible to imagine moving out. Moreover, life with the Tripathis lent me a particular vantage point on the ordinary and extraordinary events in Varanasi that was especially valuable for establishing a broader under-standing of the materiality and meaning, within which Madanpura and Varanasi's Muslims were located. Like Pinku, the Tripathis, as far as I could establish, had no significant relations with Muslims in the city or outside. I recall that whilst travelling by rickshaw one day with the eldest daughter, then ten years old, she had earnestly pointed out a lady in a *burqa,* before turning to me and explaining that the lady was a Muslim and that she ate meat. For this young girl, the notion of difference was patently understood not only in terms of dress, but also around food consumption. Evenings at the Tripathis' home were filled with discussions, often political in nature and covering local and national affairs. It was common for attitudes and opinions about Muslims to be evoked, sometimes reflecting upon the desperate situation of Varanasi's predominantly Muslim weavers following reports of suicides or weavers' selling body organs for cash payments. But more often, conversations about Muslims would express unflinching views about their anti-national sentiments, the propensity to have "too many" children and the oppressively patriarchal structure of Islam.

The family and my other Hindu friends looked upon Madanpura with a mixture of fear and intrigue. No one had the need or the desire to explore the lanes leading off the main high street, so instead, most interpreted the neigh-borhood through rumors that circulated about the presence of tunnels running under Madanpura (see Raman 2010) or how local madrasas indoctrinated future Islamist terrorists. During the month of *Ramzān,* sweet *bakha khani* was freshly baked in mobile clay ovens, ready to be eaten at sunset with the break of fast. With its syrupy taste I thought Pintu and Prabhudatt would like the sweet, and since it was a "Muslim" specialty I was sure they had not tried it, so took a couple home one day. It was therefore intriguing to witness Pintu and Prabhudatt's initial trepidation on hearing about the origin and tradition of the sweet and their subsequent caution in tasting it. Following which, there was a sense of amusement coupled with excitement that an otherwise unfamiliar aspect of "Muslim" culture had become known and tasted, although there was little indication that the food would be eaten again. These kinds of experiences, away from the immediate field, were really important in shaping my wider understanding of the city, its culture and the nature of day-to-day encounters between Hindus and Muslims.

In different ways I became conscious that the act of fieldwork itself involved challenging assumptions about the "self" and "Other" within the city. In the initial days and weeks of fieldwork, Pinku and I no doubt elicited a great deal of curiosity, as well as suspicion, from local Muslim residents who observed myself, a young white female scholar, in the company of a young Banarasi male, frequently wandering through the lanes. Our early objectives were simple; to gain an insight into everyday practices of work and the channels through which inter-community relations were forged in the locality. Opening conversations with willing weavers, curious fabric dyers and polite shop-keepers became the foundations through which stronger relationships developed and subsequent contacts emerged.

Over 200 people were interviewed (142 Muslim, 46 Hindu). These were both semi-structured interviews, covering a range of topics concerning education, work life, family context, neighborhood and intercommunity rela-tions, political opinions and participation, engagement with media and the police. Unstructured interviews, better understood as "conversation[s] with a purpose" (Burges 1984, p. 102) granted me the latitude to introduce topical matters of research and probe beyond the answers. This also permitted the chance to clarify and elaborate on answers, modify the interview as new information emerged and situate the meaning of statements within their cultural contexts (May 2001, p. 123). During the course of research, 14 key informants were established, representing a cross-section of socioeconomic and political backgrounds in the neighborhood. It was through ongoing conversations with this group that some of the richest insights were generated into the everyday politics of peace and citizenship in the city.

On returning from the field, detailed notes were recorded daily and reflected upon in order to understand daily practices of individuals in their own terms (Megoran 2006). Events and details were introduced and/or corroborated through meetings with journalists, academics and senior police officials in the city, as well as past and present newspaper sources. Given the potentially sensitive nature of information exchanged, care was taken in the field to be especially discreet (see Dowler 2001). Whilst I endeavored not to be a burden upon those I met with and encountered in the neighborhood and market, the part-teasing, part-serious quip "so how much longer until we can get rid of you then?" made by an acquaintance in the days leading up to my final departure, served to remind me how my presence, alongside a research assistant, might nonetheless be an inconvenience for some.

Working with research assistants proved to be a rich and valuable experience in Varanasi, yet little is typically documented about the process of working with someone else in the field and its impact upon the research findings (see Wolf 1996; for a prominent exception see Berreman 1963). Some perceive that the credibility of their findings may be undermined or called into question where research is co-produced and the researcher's "outsider" status is at once evident (Borchgrevink 2003). Drawing on the assistance of Pinku for a large part of the fieldwork presented both opportunities and constraints for research, particularly with respect to matters of access, communication and maintaining relations. I paid Pinku a monthly salary, determined by his going rate at the time. We had a verbal contract that the salary involved a five-day working week, and I covered travel and snacks whilst we were out. In reality, working hours sometimes extended beyond the informal agreement which I tried to recognize through additional remuneration. Given his local knowledge and self-assurance, working with Pinku lent me a tremendous degree of freedom and confidence to move around the city and its neighborhoods. Undoubtedly Pinku's Brahmin Hindu identity, which was not always initially apparent, did shape the contours of our encounters with interlocutors, who were sometimes hesitant about expressing their views. But Pinku was sensitive to local conditions and perceptions and was also savvy in foregrounding points of commonality that went beyond religious identity, notably his appreciation for Banarsipan culture and mutual acquaintances. So, whilst some individuals declined to speak freely, if at all, about their everyday lives in the city, with many more we developed a sense of trust; over the course of time they became rather candid and unflinching in their opinions.

As a Western, unmarried female carrying out research in conspicuously male public spaces, I had to navigate a particular kind of gender politics. Working with Pinku also afforded a very critical role in granting me access, security and legitimacy within public settings such as teashops but also religious processions and celebrations where large crowds gathered. Varanasi attracts tourists and backpackers from around the world so local residents were well accustomed to seeing foreigners "hanging out" at tea stalls and staying in local guesthouses. There is, nonetheless, a curious fascination with foreigners, best illustrated by the disproportionate number of stories and photos of foreigners, particularly women, in the local print media. To overcome popular perceptions of Western women as promiscuous, and to seek some kind of respect and acceptance, I tried to appear inconspicuous by conforming as far as possible to the gender norms, to the point that my "acquired" modesty became "natural" (Abu-Lughod 1986; Schenk-

Sandbergen 1992). And so I learned to lower my gaze, avoid eye contact with unknown men, and to modify my dress, always wearing *shalvār kamēz* and *dupattā* and covering my hair in Muslim neighborhoods. So keen was I to blend in that remarks from my respondents that I looked "just like a Muslim girl" were taken as compliments. However, I did not go so far as to lie about my unmarried status, as some have done (Katz 1994; Wolf 1996), instead I found it interesting to challenge popular presumptions and build rapport through sharing aspects of my cultural background and experience. To an extent I did overcome suspicious attitudes about foreign women, and in some situations, I earned the respect of male respondents, who, over time introduced me to the female members of their families and invited me to family celebrations. So, as I moved between public spaces and more private female domains fairly easily I was intensely aware of the "role flexibility" (Papanek 1964) granted to me as a Western non-Muslim woman in the city.

Given my unmistaken appearance as a foreigner I was conscious to initiate conversations and demonstrate my understanding of Hindi/Urdu whenever I could. More often, Pinku would follow up with lines of enquiry that we had discussed in advance. So that I was included in the conversation and able to respond to particular topics as they arose, Pinku translated discussions on the spot, being sure to highlight key words or turn of phrase used, all of which I recorded in a notebook. This practice became more difficult where informants talked at length, or appeared frustrated by the breaks between questions, which enabled Pinku to keep me up to speed, but sometimes left informants feeling bored and sidelined. Where translating *in situ* became tricky Pinku kept translation to a minimum and instead we would debrief immediately after the interview to make sure that as much detail as possible was recorded. Only occasionally did I use a digital voice recorder, for instance when in more formal interview situations with government officials or journalists, but otherwise the conversation flowed more easily without and note-taking proved to be an effective, comprehensive method for recording the conversations.

Central to the process of conducting fieldwork was the art of managing and maintaining relations both with my informants, and with Pinku and other research assistants. As a first-time employer I felt a strong responsibility to be open and fair about the conditions and expectations of work in field. By and large my relationship with Pinku was uncomplicated and we developed a good understanding of each other's expectations. In the early days when I was eager to be in the field all the time I found it frustrating that Pinku would sometimes have to attend to other work at short notice. However, these unexpected breaks allowed me to consolidate my findings and also make independent trips to the neighborhood, as well as meet with journalists, academics and other city officials.

Over time we developed good relationships with a number of informants, who each placed different expectations upon when, where and how often they wanted to see me and Pinku, as well as what favors I might be able to influence. I always sought to bring back small gifts from the United Kingdom (UK), and buy saris from weavers and traders whenever appropriate. However, I often felt uncomfortable about not being able to satisfy requests to bring back digital cameras from the UK, facilitate business contacts in London or take an informant's son to London for work. Jokes were often made about my role as a spy, after all why else, people wondered, would the British taxpayer be funding research in Varanasi, and amongst Muslim weavers? When informants soon realized that my position as a doctoral researcher did not endow me with a great deal of influence, attention turned to Pinku and his connections. This was relatively unproblematic; for instance, in the lead up to elections Pinku was able to impart local information about the mood of voters in south Varanasi, or introduce sari designers to a local bookkeeper and his collection of books on textiles. However, on returning to the city after a return trip to the UK, it transpired that one of our interlocutors had visited Pinku's house to ask for a "loan," which Pinku had felt compelled to meet. I was dismayed that he had been approached in this way, and whilst seeking to compensate him financially, felt really uncomfortable that my research had put him in such a position. Even now as I draft the manuscript for this book I am aware that the legacy of our fieldwork together has a more proximal and costly impact for Pinku as he continues to bump into informants in the course of his daily life.

Presenting the argument ostensibly from a Muslim perspective may attract questions concerning the balance of opinion and insight imparted in this book; however, such partiality also brings a particular value. The decision to present the story from a "Muslim" angle was driven by, first, the relative paucity of studies on Muslim agency within intercommunity settings in India and, secondly, the practicalities of fieldwork, which limited the potential for moving between Muslim and Hindu communities, whilst developing and maintaining a level of trust necessary for producing the research.

This chapter has provided the wider context within which fieldwork for this book was conducted, and through which Muslim experiences of citizenship and practices of everyday peace were embedded. Moving between different sites, the nation, the state, the city, market and neighborhood I have attempted to weave a story across scale of the complex relationship between practices and narratives of coexistence, articulations of tension and at times violence, and ongoing material realities of inequality and consequently, unequal citizenship for India's Muslims. Situated within this wider context, the following chapters draw on empirical insights generated through different urban sites and events to show how a scalar politics of peace is reproduced in dialogue with articulations of citizenship. The next chapter focuses on the aftermath of terrorist attacks in Varanasi which presents the opportunity to make peace as process visible, through particular sites and bodies in the city.

Notes

1   From Independence until 1966 Varanasi witnessed no incidents of interreligious violence. Between 1966 and 1991 twelve separate incidences of Hindu–Muslim riots were recorded (in 1967, 1968, 1972, 1977, 1979, 1985, 1986, 1989, 1990 and 1991–2).

2   In 2002 the filming of *Water* by Deepa Mehta was forcibly halted by local personalities and politicians on the Hindu right. The film plot describes the plight of Hindu widows condemned to poverty at a temple in Varanasi in the 1930s. As such, it critiques social restrictions and oppression suffered by India's Hindu widows. Although the script was approved by a national censor board, individuals and groups closely associated with the RSS claimed that the film besmirched India and constituted part of an organized plot by the Christian church against Hinduism. As filming got under way at Assi *ghāt*, a mob of Hindu fundamentalists tore down the set. The violent objections caused the UP government to retract its permission to film, fearing the escalation of violence in the city (for an account of the events see Saltzman 2006).

3   A small number of lower-caste, SC and OBC Hindus are also involved in weaving and other production activities within these Muslim neighborhoods. But, in the rural localities surrounding Varanasi this number increases to almost 50 percent Hindu SC and OBC (see Srivastava 2009).

4   Muharram is the first month of the Muslim calendar and considered to be a month of remembrance. During this time Muslims mourn the martyrdom of the Prophet Mohammed's grandson. In Varanasi the event is publicly commemorated by Shia Muslims in particular who process through the streets carrying decorated tombs (*tāziyā*), whilst young males perform acts of self-flagellation.

Chapter Three
Making Peace Visible in the Aftermath of Terrorist Attacks

On Tuesday, March 7, 2006, just one week before the Hindu festival of Holi, Varanasi was the target of two terrorist attacks. The first bomb exploded at the Sankat Mochan temple as evening prayer was due to commence; 21 people were killed (*Times of India* March 8, 2006). Two others died 15 minutes later in two further bomb blasts at the Varanasi Cantonment railway station, one near the enquiry office and another on Platform 1, from which the Shivganga Express was about to depart for New Delhi (*Times of India* March 8, 2006). In addition to the 23 people killed in the attack, over a hundred were injured (Pradhan 2006).

This chapter places peace in Varanasi and points to the situation of the city's Muslim population by provoking some of the larger questions examined throughout the book concerning the ways that peace is situated, constructed and reproduced. And, in whose image peace is imagined. It attempts to make peace visible as a process by documenting the events that influenced the active maintenance of intercommunity peace after the city became the target of bomb blasts.

It might seem incongruous to emphasize the importance of understanding everyday peace in light of terrorist attacks on the city. On the contrary, that

Everyday Peace?: Politics, Citizenship and Muslim Lives in India, First Edition. Philippa Williams.
© 2015 John Wiley & Sons, Ltd. Published 2015 by John Wiley & Sons, Ltd.

this incident of terrorism, arguably designed to destabilize everyday peace between Hindu and Muslim communities, did not result in violence provides us with a valuable insight into intercommunity relations in the city. The exogenous shock of the terrorist attacks effectively lifted the lid on everyday Hindu–Muslim relations as the city became a site of anticipated communal tension. The underlying question on everyone's minds was: Would the terrorist attacks instigate further violence, or would everyday peace be maintained?

One of the challenges of studying peace beyond "postconflict" zones and in more "undramatic" contexts is knowing what peace looks like. In this chapter I argue that the aftermath of a terrorist attack provides useful analytical purchase for making peace visible in the city. It opens up the lines of sight to the everyday sites, scales, bodies and narratives through which peace is negotiated. And, it illustrates how peace is a process that is contingent on the legitimacy of certain agencies and narratives in the presence of violence and the threat of further conflict. The chapter begins by outlining the series of actions that followed the bomb attacks and the narratives surrounding the incidents that were interpreted not only as attacks on a city, but also on the Indian nation. The second section examines the actions of national and local political party members as they sought to influence the maintenance of peace but ultimately failed to win the trust of local residents upon whom their legitimacy was contingent. The third section shifts attention to the everyday shared spaces that existed outside of the state and were important contexts and producers of peace, most notably the silk sari market. In the fourth section I argue that local religious actors played critical roles in brokering peace by embodying forms of peaceful sociality between Hindu and Muslim communities. Answering why this was possible, the chapter opens up the question of relative agency and legitimacy and shows how local agency was derived through vernacular expressions of "peace talk" and spatial practices of everyday peace that had a particular situated geography. The presence of strong intercommunity networks in the silk sari industry constituted a key motivation for maintaining peace as well as the pathways and peaceful imaginaries to do so.

Attacks on the City, and the Nation

The terrorist attacks were intended to hit at the heart of the Indian nation. As a seat of orthodox Hinduism, Varanasi occupies a significant symbolic site within the Hindu imaginary and has long been venerated as one of the seven cities of Hindu pilgrimage. Most of the Hindu deities are enshrined in their own space, and manifold myths record their origins (Dalmia 1997, p. 50). Since the Ayodhya issue, Varanasi, along with Mathura, has been posited as a sensitive site for the politicization of Hindu–Muslim relations. The Kashi Vishwanath temple and Gyan Vapi mosque complex has been the target of rhetoric by the Vishwa Hindu Parishad (VHP). Party members have threatened to demolish the mosque in a move resembling the destruction of the Babri mosque in Ayodhya. The terrorist attack did not strike the heavily guarded temple and mosque complex in the narrow lanes of the city's busy Chowk area. Instead, it targeted the popular Sankat Mochan Hanuman temple, which is situated in the newer middle class neighborhood of Lanka in south Varanasi.

On Tuesdays and Saturdays the temple is particularly crowded. The Tuesday evening in question was no exception as devotees – men, women, children and students – had thronged to the temple. The bomb was planted in a pressure cooker disguised as a gift for one of the wedding parties in the temple. Over a thousand people congregated outside the temple gate; some were screaming and shouting, or crying in anger and weeping. Casualties were rushed to the nearby hospital assisted by volunteers, students and auto- and bicycle-rickshaw wallahs.

The *Mahant* (chief priest) of Sankat Mochan temple believed that normal activities should be resumed as soon as possible in an effort to calm the community. He expressed alarm at the lack of police presence after the attack. In an atmosphere of heightened emotions, the *Mahant* feared a violent backlash, given that the senior superintendent of police (SSP) and the district magistrate (DM) were both away on leave.

With the media competing to break the news, the *Mahant* found himself in the spotlight; his interviews were broadcast worldwide. He reiterated the same message, that *Shanti kisi bhi hāl mein banaye rakhna hai* (peace must be maintained at all costs). By 9.30 p.m. the *Mahant* was thankful that *ārti* (an evening prayer ritual) could commence. In the meantime, the *Mahant* received a telephone call announcing the arrival from Delhi of Sonia Gandhi, the leader of the Congress Party, accompanied by the Union home minister. "I couldn't believe it, that she [Congress leader] of all people was coming… and at night … I was really very moved and very touched by her visit." After offering *darshan* (seeing the divine image) to Lord Hanuman, the leaders visited the hospital to meet the victims of the blasts.

The following day, party political leaders and party members, religious figureheads and celebrities visited Varanasi and the temple to demonstrate their sympathies and condemnation of the terrorist attacks. The Congress Party described the attacks as a "senseless act of violence" and appealed for peace and harmony. The general public were cautious about naming terrorist organizations and especially about blaming the Muslim community. When I presented this subject to the *Mahant*, he responded by saying:

> There was no information for me, so I couldn't blame X, Y, Z. The news that it was a terrorist attack spread by word of mouth... people have this belief that all terrorists attacks are conducted by one particular community so the fingers were pointed towards them. But I did not say this; even today I don't say this.

A similar prudence was evident amongst Hindu friends. People did not openly volunteer opinions, but when prompted would suggest that perhaps the terrorists were members of the "special caste" or "special religion," thereby implying that members of the Muslim community were involved.

The visits by political party members were widely received with skepticism and regarded as political opportunism rather than expressions of genuine sympathy. A history of poor administration in UP framed many of the responses as a local Hindu businessman described:

> They [the politicians] only came here to show their sympathy because in six months there is a state election coming up. We don't want them to come here. We don't want Sonia Gandhi or L.K. Advani, or an X, Y, Z political party to come and put cream on our wound. We want something to be done about the state of our utilities and services.

Both Hindu and Muslim leaders were determined to minimize the potential for conflict between their communities. When a BJP leader attempted to stage a *dharna* (peaceful sit-in protest) in the temple precinct, his political gesture was ill received by the *Mahant* who firmly insisted that political activities were not welcome in the temple. In line with the *Mahant*'s insistence that the incident of the bomb blast would not be exploited for political gain, the *Mufti* of Varanasi spoke to the community: "My appeal is also to all sections of the people and political leaders to refrain from making our city the boxing arena for settling political scores and deriving political mileage from such tragic incidents." At the national level, the home minister in New Delhi immediately raised the national terror alert to red. Sensitive areas were policed including high-profile temples in New Delhi, the Raghunath temple in Jammu, the Akshardham temple in Gujarat, and several temples in Mumbai as well as in the holy cities of Haridwar and Somnath. Hundreds of additional security forces were deployed in UP (*Times of India* March 8, 2006).

On March 8, more policemen were drafted into the city and visibly deployed on the streets, particularly in the sensitive regions of Kamovesh, Jangan Bari, Madanpura, Sonanpura, Bhelupura and Reori Talab (*Hindustan* March 8, 2006). Directly after the blast and throughout the following day the city was silent. A school principal described how she and her mother spent the entire day cooped up in their flat waiting, hoping that everything would remain calm in the city. Varanasi adopted an unofficial curfew. As news of the terrorist attacks spread across the city on the evening of the blasts, businesses in both Hindu and Muslim localities promptly pulled down their shutters in acts of sympathy and anxiety. A middle aged Muslim *Ansāri* friend and father recalled how on hearing of the bomb blasts he immediately thought to with-draw cash from the automatic cash machine on the main high street so that, should violence ensue and/or an official curfew be implemented, he would have money available to look after his family. Leaders from the Muslim communities in Varanasi unequivocally condemned the blasts and expressed their community's grief and hurt at the attack. The *Mufti* of Varanasi voiced his concern regarding the possibility of a Hindu backlash. Directly after the blasts, he appealed to his community to maintain the peace, pressing them to ignore rumors, maintain their "*bhaī-bhaī*" or "brotherhood" relations, and remain calm.

The media coverage of the bomb blasts was extensive. The vernacular and English media were persistent in running the story of "communal harmony" and resilience in Varanasi. Just hours after the blast, satellite television channels hosted debates with leaders from different religious communities to demonstrate the unity of Varanasi. Newsprint media ran the headlines: "A day after Varanasi blasts, police in Delhi and Lucknow crack down on terrorists even as the holy city bounces back to life" (*Times of India* March 9, 2006) and "A resilient Varanasi bounces back" (*Times of India* March 10, 2006). High-profile personalities were featured; Bismallah Khan, the famous Muslim *shenai* player of Varanasi, appealed to the people of Kashi to keep the peace and maintain the *Gangā Jamuni tahzēb/sanskritī* of the city (*Hindustan* 8 March 2006). *Gangā Jamuni tahzēb/sanskritī* (*tahzēb* Urdu. "manners" or way of life. *Sanskritī* Hindi. "culture") refers to the confluence of the River Ganges and River Jamuna in the region, the former is associated with Hindu rituals and the latter with Muslim connotations. More broadly this narrative referred to a history of cultural collaboration between Muslims and Hindus with regards to their artistic, literary and musical production in the region.

As Wednesday, March 8, passed without incident, the DM reopened schools and colleges the following day, and life quickly returned to the streets. Reflecting on the situation weeks later, the sentiment conveyed in the press, and imparted from friends and interviewees, was that of relief, underpinned by pride for the manner in which the city had responded. It was clear to all that this incident could have provided the spark for com-munal riots in the city. One housewife remarked that, "It was such a good thing, such an impressive thing that there was no trouble in the city after the blasts … especially when we see such conflict between Hindus and Muslims in Aligarh and Meerut."

Even whilst Varanasi is widely imagined both in local and national spheres as a peaceful city, it has not been exempt from occasional incidences of communal clashes, often attributed to a dispute over a local religious structure or procession (Engineer 1995). In November 1991, the *Ramjanmabhoomi* controversy in Ayodhya provided the impetus for riots, as in many cities across UP and India. The riots in Varanasi however, were far less extensive than elsewhere in the state. Yet, in March 2006 there was genuine reason to believe that the terrorist attacks on the temple and railway station could inspire further, more extensive intercommunity conflict. During this period, communal tensions in Uttar Pradesh were widely perceived to be on the increase. In October 2005, communal riots broke out in Mau, another center of the handloom industry 90 kilometers from Varanasi, leading one writer to conclude that "eastern UP is sitting on the mouth of a communal volcano" (Sikand 2005). On March 5, 2006, four days before the Varanasi terrorist attacks, violence between Hindus and Muslims erupted in Lucknow during protests staged by Muslim organizations against the imminent visit of United States President George W. Bush. Party politics in UP capitalized on the events. The Samajwadi Party framed the incident in Lucknow and the Indian government's vote against Iran over nuclear armament as anti-Muslim moves. Simultaneously, the Bharatiya Janata Party accused the ruling coalition party at the center, the Congress-led United Progressive Alliance, and the Samajwadi Party in the state, of minority appeasement agendas. In the aftermath of the Varanasi bomb blasts, we saw a further terrorist attack on the Jama Masjid in Delhi, communal riots took place in Vadodara, Gujarat, and communal clashes occurred in Aligarh in May 2006 and Gorakhpur in January 2007.

The fact that peace was maintained in Varanasi in light of the political climate in Uttar Pradesh appears all the more interesting. It is possible that central to the explanation is the mode of disturbance itself - the terrorist attack. A cursory survey of the terrorist attacks in the Indian subcontinent between 2002 and 2006 reveals that none of these provided the impetus for communal-based violence. Perhaps terrorist attacks do not destabilize community relations to the same extent that a dispute over the route of a religious procession might. It is possible that, because the perpetrators were framed as external elements, local communities regarded the "Other" as a foreigner rather than their own neighbor, even uniting communities in opposition to an external "Other." The terrorist attacks in Varanasi, nonetheless, created an atmosphere of anticipated communal tension. The bomb blast at a site of religious worship could have been translated as an attack on an entire community, or even the entire "Hindu nation." From this perspective the terrorist attacks provide a functional moment in which peace is made visible, where relations between Hindus and Muslims are exposed, and where the capacity of different actors and mechanisms to successfully resolve tensions through different pathways to peace can be examined. The next sections, therefore, seek to excavate the different but interrelated mechanisms and agencies which functioned towards peace, in particular considering the importance of the state and political actors, shared intercommunity spaces and local actors. Why these actors acted as they did, and the extent to which their actions resonated with city inhabitants, must be seen as contingent on their legitimacy within real and imagined shared spaces of Varanasi, and India more widely.

The State, Party Politics and Mechanisms for Peace

I conceptualize the state both as the system that encompasses government and legal institutions, and as the idea of a "sovereign entity set apart from society by an internal boundary that seems to be as real as its external boundary" (Fuller and Harriss 2001, p. 23). Despite the difficulty in determining the boundary between state and society, it remains conceptually significant (Mitchell 1999, p. 78), not least because modern politics concerns the production and reproduction of that boundary. The argument set out in this chapter rejects the Weberian structural definition of the state and instead subscribes to a multifaceted understanding of the state, from the perspective of everyday reality and prosaic practices (see Painter 2006). The state is not a single organization distinct from society. Rather, it is composed of a multitude of bodies acting sometimes in coordination with each other, but more often exhibiting individual agencies that may conflict, contradict and compete with others (Fuller and Harriss 2001, p. 22), with unforeseen, occasionally creative consequences for the state and society (Corbridge et al. 2005; see also Donegan 2011). Given this interpretation, it reinforces the importance of focusing on the lived experience of the state, both as a collection of agencies and as an idea (see Williams et al. 2011). Some state and central governments have acted decisively in dealing with potential riot situations and have, at other times, not acted at all or have been ineffective (Brass 2003, p. 374). Evidence suggests that even the weakest state governments still possess the necessary capacity to control and contain riots when they develop and possibly to prevent them occurring (see Brass 1997; Wilkinson 2004). The efficacy of state agencies during periods of communal tension is thus highly dependent on the political and social context.

The decision by the Congress leader and UPA home minister to visit Varanasi in the early hours of the morning following the attacks, represents decisive action by central government to ensure that communal harmony was maintained in Varanasi and across the nation. Besides demonstrating solidarity with victims of the blasts, the central government strongly condemned the terrorist attacks and called for the maintenance of peace in the city. The political climate in UP contributed towards the UPA's swift action. The government's perceived recent conciliation towards the Muslim community over the issue of protests against the publications of cartoons in Denmark depicting the Prophet Mohammad and the protests by Muslim organizations against the visit of the U.S. President had already exacerbated tensions. The Congress, particularly sensitive to the potential of a Hindu backlash, acted to minimize the potential space for agents to capitalize on the situation and polarize communities along communal lines. The reactions by state party political leaders reinforced the moderate message of the center by universally condemning the blasts and appealing for restraint and calm.

How the government's rhetoric of peace and restraint fed into the actions of the police administration is less evident. In the immediate aftermath, law and order occupies an interesting place in the narrative by virtue of its low profile. Neither the SSP nor the DM was present in the city at the time of the attacks. Some reports cite that the police arrived 20 minutes after the first bomb exploded at Sankat Mochan temple (*Times of India* 10 March 2006). Under these circumstances, it was necessary for the temple administration and arguably the public to introduce their own strategies to manage the period after the attack.

Following the attacks, police presence was heightened in the city, with security reinforced by the Provincial Armed Constabulary (PAC) forces at designated "trouble spots," near Hindu and Muslim community boundaries. By March 10, there were no untoward incidents reported anywhere in the district (*Times of India* March 10, 2006). In part this may be down to the high profile of the police and possibly even their quick reaction at the sign of trouble; certainly, the protest against the UP chief minister's visit was dispersed after a police *lathi* (stick) charge (CNN IBN report March 8, 2006).

Wilkinson's (2004) study puts forward the supposition that during a period of potential violence the mode of reaction by a government's law and order machinery is determined by a government's electoral interests. In a state like UP, where party political competition has progressively fragmented the electorate, there is the possibility that the government would act to protect minorities because it is in its electoral interests to do so. As Wilkinson (2004, p. 238) argues, "Politicians in highly fractionalized systems must provide security to minorities in order to retain their electoral support today and preserve the option of forming coalitions with minority-supported parties tomorrow." To what extent electoral calculations can be attributed to the prevention of violence is debatable; within the same state and same electoral timeframe, interethnic violence was not prevented in Mau, Aligarh and Meerut.

This fact suggests the presence of other agencies operating outside the purview of government institutions that had an interest in preserving peace in Varanasi and the capacity to do so. While the BJP joined state parties in calling for calm, it did not hesitate in directing the blame towards Pakistan and by association towards the Muslim community. The intention of the party's leaders was to maximize the potential political profit gained by mobilizing supporters around the terrorist attacks. BJP chiefs announced the launch of twin "national integration *yātrā*s (journeys)" (*Times of India*, March 11, 2006). Protests were conducted across the city opposing the terrorist attacks and the failure of the state government to avert them. A BJP leader attempted to stage a *dharna* in the temple. Notably, these initiatives, to politicize the community along communal lines, failed to capture the public imagination and inspire renewed support for the Hindu right-wing party.

If the BJP initiatives were designed to inculcate communal tensions for political gain as they have successfully done in the past, why did the plan fail in this particular context? In 1992 their *rath yātrā* (chariot journey) proved to be a powerful instrument in mobilizing the electorate around the issue of the Babri mosque. Similarly, Hindu right politicians have been blamed for inspiring communal violence. Brass describes the role of an "institutionalized riot system" to explain the "production of Hindu–Muslim violence" and cites "communal mobilizers" (the professional politicians and activists in the BJP and RSS and associated organizations) as those destructive agents who take advantage of communal tension to plant the seeds for a communal backlash (see Brass 2003, pp. 377–9). Riots are produced by precipitating events, such as the killing of a politician, the theft of a religious idol, or an attack on a religious place of worship. "One reaction then leads to another, generating a chain, which if not immediately contained will lead to a major conflagration" (Brass 1997, p. 257). In this case, the terrorist attacks in Varanasi did precipitate reactions by the BJP and members of the Sangh Parivar, stimulating the production of a first link in the chain. Subsequent links failed to materialize. Why sufficient momentum could not be generated to enact a second link in the chain demands several explanations. First, the BJP was struggling to unite its members following its defeat in the 2004 Lok Sabha elections. The political fatigue and infighting of party members in association with a lack of clarity in its political direction simply failed to inspire its electoral base. Even the announcement of the *rath yatra* was characterized by discord amongst party members disputing the timing and appropriateness of the strategy (see Mukerji 2006).

Second, Brass (1997, p. 258) suggests that when "an effort to create the conditions for a riot by an identifiable political party or organization such as the BJP from which they stand to benefit becomes too 'transparent'… people do not want to be manipulated." This attitude was conveyed by my respondents. Not only were the actions of the BJP transparent, but there was a sense of fatigue towards their style of political mobilization. Moreover, the nature of the provocation on which party members attempted to capitalize, the terrorist attacks, was regarded as inappropriate. In the first instance, people wanted to see *real* empathy from the politicians and *real* help for the victims and their families; instead, they were incensed by the blatant political opportunism demonstrated. The style of the terrorist attacks meant the perpetrators were not visible members of the local community. Although not confirmed at the time, it was widely believed to have been conducted by external elements, Islamic fundamentalists from outside India. One milkman told me: "We think terrorists are to blame. We can't blame any special caste, because terrorists have no caste, we can't blame Muslim or Hindu or Christian. I only know that terrorists are responsible. Terrorism itself is a religion." The clear differentiation made by Hindu respondents between terrorists and Muslims informs the other half of the explanation for why the BJP failed to inspire support around this issue. Central to their mobilization campaign is the threat of the Muslim "Other"; however, in this instance the "Other" was a terrorist, perceived as distinct from the local Muslim community. Both Hindus and Muslims were the possible targets of the attack at the Cantonment railway station, and both communities were similarly angered and saddened by the blasts. Concurrently, blame was not directed towards the local Muslim community.

Distancing the State?

In this context actions and narratives by members of society conveyed the belief that state agencies and formal political parties were best kept at a distance from everyday civil society, especially where there was potential for events to become politicized. When I explored the impact of the state on the lives of my respondents, they demonstrated an element of wizened cynicism towards the response of state agencies. It was evident that both the Hindu and Muslim communities in Varanasi did not rely on the local government and administration for guidance and support during this tense period, but instead looked to their own local communities. A printing press owner remarked how the "administration and politicians didn't do anything [to maintain the peace] … it was the people who demonstrated their morality and humility not the politicians." This opinion was corroborated by a local doctor: "The community led the way and showed its strength, not the political leaders."

This sense of distrust evidently extends beyond the contemporary state government to encompass feelings of frustration and skepticism towards the actions of politicians in general. Political parties have repeatedly failed to provide the basic utilities in Varanasi, creating a very disappointed electorate. The common sentiment is that politicians act purely in their own interests to win votes, while the people are left to fend for themselves. In association with this disillusionment was a palpable desire for politics to be kept separate from everyday community relations. Interference by politicians and the police was believed to exacerbate tensions and sometimes even stimulate trouble, as a young Muslim weaver informed me: "If the administration wants to, they can create a riot like in 1992 … Then a nervous atmosphere was created and so there was a riot. But when it's left to the people we don't want to riot, we never want to fight with each other." When it comes to a question of the preservation of law and order, the *Mufti* of Varanasi further reinforced the importance of distancing the state by commenting:

> Everyone is of the same opinion that only we can save ourselves. Not the police, not any security, not the administration. We don't believe it's possible to rely on external structures … The religious leaders and the general public, these are the only two responsible for maintaining peace. The government has nothing to do with it.

These findings resonate with observations made by Shah (2007, 2010) in her study of the Mundas in Jharkhand. Often depicted as poor tribals, this group of people experienced the state as exploitative and at odds with their political and ethical ideals, which consequently informed their desire to actively keep the state out of their matters. Similarly, history had revealed the provocative and often partisan role of the state in intercommunity matters. Residents in Varanasi articulated a strong belief that potential intercommunity tensions were therefore best resolved without the interference of the state. Community leaders positively encouraged a distance between politicians and the affairs of local community relations in order to preserve the peace. But, what aspects of societal interaction in Varanasi enabled community leaders to work effectively in alleviating potential differences and why were certain community leaders eager, and moreover, able to keep the peace?

Everyday Shared Spaces

The emphasis placed on the importance of community leaders and actors operating "outside" the state in explaining the maintenance of peace was significant. Such findings strike a chord with the study conducted by Varshney

(2002) which orients civil society towards the center of our understanding of the occurrence of communal violence in some timeframes and localities but not others. Where Varshney's argument emphasizes the importance of inter-communal ties within this shared space in facilitating peace, my argument goes further. By exploring the characteristics of Varanasi's everyday shared spaces, I suggest that it is not just the structure of society, but also the actions of agents, both enabled and constrained by that social and political context, which are particularly crucial in explaining the preservation of communal harmony.

Respondents repeatedly pointed to the strength of pre-existing social relations between the Hindu and Muslim communities in order to explain why Varanasi had not been party to episodes of communal violence following the terrorist attacks. A middle-aged Muslim factory owner told me that:

> We were utterly worried that there would be trouble after the blasts. The population of Varanasi are thankful because the brotherly atmosphere in the city has been maintained ... The feeling of brotherhood, our mutual under-standing, mutually helpful relations and sense of togetherness are all responsible for the peace.

Similarly, this sense of mutual respect between different communities was articulated by a Hindu shop owner from Lanka:

> Banarasi people want to enjoy life, they don't want to run after money ... OK, now the generation is changing, everything is changing, but in spite of that they have close relationships with each other ... in spite of the different religions they follow, or what business they are doing, which layer of society they belong to, they love each other.

What unites Hindus and Muslims in Varanasi? Is there something distinct about the relations in this city compared with those in other north Indian cities? The views of a local journalist were illustrative as he emphatically stated that:

> The main reason for peace in the city is because of the *Gangā Jamuni sanskritī* ... people think that we should live together and overcome this problem [terror-ist attacks]. If you go anywhere like Delhi, Bombay, Calcutta people are not ready to help you, but here because of the Banarasi culture people are ready to help you.

Such sentiments reflect a sense of situated everyday cosmopolitanism in which local Banarasis were able to negotiate difference and find common ways of relating through the city's shared vernacular cultures. Drawing on fieldwork conducted in the 1980s on the cultural systems of artisans in Varanasi, Kumar (1988) points to the cultural factors that unite Hindus and Muslims in the city. Respondents frequently referred to themselves as "Banarasi," and alluded to a particular style of life unique to Varanasi and one of which they are very proud. Alongside a belief in the religious significance of the city, Hindus and Muslims alike hold another set of beliefs on the virtues of the city, quite removed from the immediate world accessible to pilgrims and tourists. The Banarasi lifestyle revolves around three defining symbols of the city: the Ganges, the temples and the bazaars. While the importance of the Ganges is variously interpreted by different communities, the river as a site for boating and swimming, relaxation and contemplation, tea-drinking and chess-playing is experienced by people across the religious communities (see Kumar 1988, pp. 72–3). A local journalist pointed out that it is on the *ghats* (the steps down to the river) that the Ganga–Jamuni culture may be witnessed:

> In Banaras we have the *Gangā Jamuni sanskritī*, if you want to see this you just have to go to Dashaswamedh Ghat or Assi Ghat, people just want to live together because this is the city of *mauj musti* (carefree fun)… everyone has a love for each other in Banaras.

Kumar concludes that this cultural system is what unites Hindus and Muslims because it is not primarily divided into categories of religion or caste, but instead the most important divisions, in order of importance, appear to be men/women; upper class/lower class; educated/uneducated; Banarasi/non-Banarasi (Kumar 1988, p. 225). More importantly:

> Hindus and Muslims of the lower classes share a similar lifestyle and ideology of work, leisure and public activity. And while that of the upper classes has changed over the century, that of the lower classes has remained structurally and culturally the same. (Kumar 1988, p. 226)

This observation may still resonate with reality in parts of Varanasi. However, my strong sense was that subsequent to Kumar's research in the 1980s the religious–political climate of India had undergone a significant shift such that religious distinctions had come to matter more than Kumar's study suggests. Whilst recognizing that religion is not the only or most significant axis of difference, set in the mid 2000s this book contends that the distinctions between Hindu and Muslim self identities and collective identities were consciously negotiated and mediated, often in contradictory ways. While many of my respondents enthusiastically articulated the importance of brotherhood and shared culture between Hindu and Muslim communities, there was an intriguing disconnect between this articulation and their own direct experience of brotherhood. As a means of generating further links with Muslims, I asked predominantly Hindu respondents if they had any Muslim friends in the vicinity with whom I could meet. The response was generally muted; after taking some time to think, only a couple of respondents could point me in the vague direction of a house which they *believed* to be occupied by a Muslim family.

The ward of Nagawa is predominantly inhabited by Brahmins, businessmen, Yadavs (a Hindu caste cluster) and a large community of Bihari migrants. This would certainly explain the low level of direct social or economic interaction between Hindus and Muslims. What struck me, however, was that the perceived "brotherhood" in the city still formed such a strong discursive construct, almost universally expressed by my respondents. Furthermore, Brahmin friends would positively attribute the strength of Hindu and Muslim relations in the city to the potential for peace, alongside personal sentiments revealing an inherent disdain for Islamic culture and morals coupled with broader suspicion of the Muslim community. For these Brahmins it did not appear contradictory that direct social interaction with Muslims was not a feature of their everyday experiences, yet in explaining the wider dynamics of the city, such interactions, it seemed, were regarded as integral to its peaceful equilibrium. This presents a situation where everyday interaction between the Hindu and Muslim communities characterize certain zones of society, be they vertical or horizontal, but not others. The imaginary of Hindu–Muslim brotherhood, on the other hand, was universally appreciated.

The city's silk sari industry represents one arena in which associational integration between Hindus and Muslims was tangible. Chapter Five further explores the nature of these relations and practices of power within the silk industry; however, it is important to introduce the industry briefly here, given its role in creating the context in and through which peace was maintained after the terrorist attacks.

The silk sari industry employs an estimated 500,000 people as weavers, dyers, sari polishers and traders in and around Varanasi. Traditionally, the trading sphere has been dominated by the Hindu community, whilst Muslims make up the weaving workforce. In recent years, an increasing proportion of master-weavers from the Muslim community have become wholesale dealers (Showeb 1994). This has produced a trading environment, as a friend described, where "horizontal and vertical integration means that Hindus and Muslims are thoroughly intertwined and interdependent in the business … Even today Hindus and Muslims are great friends in the city."

Yet, to ascribe the maintenance of peace to the presence of interdependent economic relations between Hindus and Muslims in the silk industry is too simplistic an argument, particularly since economic relations often constitute sites of tension. Engineer's (1995) analysis of communal violence in Varanasi suggests that economic competition between these two communities has sometimes generated violent incidences. The challenge presented to the hegemony of the Hindu trading classes by "weavers-turned-entrepreneurs" was a contributing factor to the occasional communal clashes between 1970 and 1990 (Engineer 1995, pp. 197–206). Taking into account the contemporary social and economic context of the silk industry, I suggest that the dynamics of the silk industry not only facilitated, but also influenced efforts to maintain peace in the city.

Over the last 10 years, the Varanasi silk industry has undergone a period of decline. Increasing market competition from China following the 1996 economic liberalization policies has progressively undercut the city's silk market. This, coupled with a growth of the power-loom sector in Surat and Bangalore, and the changing demands of the market, has increasingly crippled the Varanasi weaver (*Times of India* September 4, 2005). The weavers hardest hit by the depression are inevitably the impoverished handloom and small-scale, predominantly Muslim, weavers, whose capacity to diversify was restricted by lack of capital. This had created a situation of high unemployment, compounding feelings of insecurity rooted in the perception of a worldwide antipathy towards the Muslim community fueled by the actions of the United States of America and the United Kingdom.

The apparent centrality of civil society in ensuring the maintenance of peace after the terrorist attacks strongly resonates with recent findings by Varshney (2002), who concludes that intercommunal interactions are the single most important factor behind the maintenance of communal harmony. While Brass (1997, 2003) and Wilkinson (2004) contend that the state has the power to prevent communal violence if it so wishes, Varshney (2002) does not agree. Instead, he argues that the effectiveness of the state response is, in part, a function of how integrated Hindu and Muslims relations are in society.

Invoking Putnam's (2000) concept of social capital provides a productive insight into these dynamics where some types of networks are valued over others. He distinguishes between social "bridging" and social "bonding." The former refers to inclusive relationships that reach across diverse social cleavages; the latter denotes exclusive relations that create strong in-group loyalty, reinforcing a homogeneous group identity, and in the process creating out-group antagonism. Varshney (2002) applies these concepts in the context of relations within and between Hindu and Muslim communities. He contends that networks of intercommunal or "bridging" relations enable tensions and conflicts to be regulated and managed, and that where such networks are missing at times of tension, violence generally results. Varshney further distinguishes between everyday intercommunal integration and associational integration. The latter is formed in an organizational setting, while the former requires no organization. Whilst both engagements promote peace, associational engagements are sturdier than everyday forms in dealing with communal tensions (Varshney 2002, pp. 9–10).

Varshney's (2002) thesis implies that everyday intercommunity relations need to be universally experienced across the city to make communication between the two communities possible during periods of tension. This assumption appears rather insufficient. In Varanasi, it is apparent that some sections of the communities engage in real everyday relations and/or associational networks, while other sections experience little or no intercommunity interaction. Take the illustrative case of Rupa, a Brahmin Hindu lady in her mid-thirties, married and living in the southerly neighborhood of Nagawa: the opportunity for her to interact directly with Muslims is a rare event. But, whether or not Hindus experienced everyday interaction with Muslims or vice versa, what is relatively universal across both communities is the discursive construction of their Hindu–Muslim brotherhood or "*bhaī-bhaī*"/ "*bhaīachārā*," a relationship that is proudly attested. This construct is in many respects an "imagined community"; the connections that bind people are not always personal ones and those who speak of its existence do not necessarily experience everyday intercommunity interactions, nor desire to. A tension exists between the notion of an "imagined community" and Varshney's (2002) thesis based on personal intercommunal interactions. In reality, the two work hand-in-hand. Whether the Hindu–Muslim brotherhood or community is something actually participated in or just imagined does not make the idea of "the community" any less real. However, as I document in Chapter Five, the discourse of "Hindu–Muslim brotherhood" is not a neutral one, but intimately connected with the political. I contend that in this circumstance it was the *belief* in a universal Hindu–Muslim brotherhood that enabled community leaders, as well as the media, to appeal to a common imagination in their calls for peace.

Local Agency, "Peace Talk" and Legitimacy

The events and narratives that unfolded in Varanasi also illustrate the importance of individual personalities in their capacity to implement mechanisms enabling the maintenance of peace. By contrast the question of agency and the actual mechanisms that facilitated the mediation of communal tensions is relatively unexplored by Varshney (2002). Where Varshney's argument emphasizes the value of intercommunity ties in facilitating peace, my argument goes further to suggest that it is also the action of particular agents within civil society which matter, whether they support or undermine inter-community ties. It is also vital to understand the ways in which these actors are enabled and constrained by their situated geographies. For instance, pre-existing intercommunal ties of engagement alone may not have produced peace in the face of efforts by the BJP to exploit the situation for political gains. The reaction of the *Mahant* was critical in this instance. The approach developed by Brass (1997, 2003) which recognizes the importance of individual actors in stimulating a chain of events that do or do not culminate in communal violence provides us with a useful framework. The visit by the Congress leaders, the calm response by the *Mahant*, and the calls for solidarity by the *Mufti* of Varanasi, as well as actions by other personalities in their respective localities, positively acted to set in motion a chain of effects that worked to minimize the impact of the shock on their communities and enable harmony to prevail.

The initial weakness of the police force in the city, but the safeguarding of peace nonetheless, points to the significance of the constructive reaction by the *Mahant* and the general public. The *Mahant*'s strong leadership, under-pinned by his religiously inclusive, secular credentials, was critical in main-taining communal harmony in the city. Directly after the bomb blast he and his colleagues were conscious that *ārti* should resume as soon as possible, projecting the idea that life can and must continue as normal. For local and national media, the temple's official message was one of peace, not anger or blame. These decisions were informed by an acute awareness that the manner of events at the temple could impact on the psyche of the Hindu community. Had the attack been construed as an assault on the whole Hindu community, this could easily have created the potential for a violent backlash.

As other party political leaders visited Varanasi in the days and weeks after the blasts, the prompt visit by the Congress leader was regarded as a poi-gnant gesture, not just for the *Mahant*, but also for members of the community near the temple. A Brahmin friend and lifelong BJP supporter even described his admiration for the Congress president. While the actions of other party political members were treated with cynicism, this leader's actions were seen as sincere and heartfelt. The evident magnetism of the Congress leader and the perceived significance of her visit hints towards the weight of personality in Indian politics, as well as the widely held regard for those in positions of power.

With his secular outlook, the *Mahant* implicitly and explicitly countered attempts by the BJP to politicize the incident. On two notable occasions he refused to acquiesce to the demands of a BJP leader. First on the matter of staging the *dharna*, and second, following the visit by a Muslim lyricist to the temple, the BJP leader's insistence that the temple should be "cleansed" with holy water from the Ganges. The *Mahant*'s firm rejection was a conscious decision to ensure that Hindu worship should not be confused with radical Hindu politics, as he made quite clear:

> We are all practising Hindus; we are not radical Hindus and so the temple is for worship... the radical Hindus, or radical Hindu parties, are exploiting the sentiments and are not helping the process... And if there is any practical problem with religious institutions, any problem created by the society today, they are not going to come to our rescue or help. They will suck the blood and go. That is what I feel.

Whilst the constructive agency of the *Mahant* certainly played an important role in minimizing potential violence and communicating a message of routine calm, it was the partnership he formed with *Mufti* that was most widely attributed to shaping the course of peace in the city (Gupta 2006). It is said that the *Mufti* and *Mahant* were introduced by one of the *Mahant's* sons in the days following the attacks to discuss how they could form a united public face for and of the city in order to allay the possibility of a backlash by the Hindu right. In the days immediately after the attacks the religious leaders gave interviews together on news programs, voiced their collective sympathies in local newspapers and were photographed together in national current affairs magazines, amongst countless other publications. In a nation that is sensitive to its religious–political differences, the image of a Hindu priest and Muslim cleric speaking on the same platform with one message is a very powerful one.

In a photograph that appeared in *Outlook*, a nationwide current affairs magazine, the two men are sitting on plastic chairs on the terrace outside the *Mahant's* house, the *Mufti* is to the left wearing a skull cap and *keffiyeh* (Arab style headdress) over his shoulder, and the *Mahant* is dressed in white *kurta-pajamas*. Both men look relaxed and to be enjoying each other's company. The strategic bringing together of bodies marked by their corporeal differences, in order to create spaces of peace, evokes some parallels with the research of Koopman (2011a) on peace accompaniment in Columbia. There, she shows how less vulnerable US citizens move alongside more vulnerable Columbian peace activists to enable situated protection from armed groups. Here, however, it is imagined spaces for peace as intercommunity solidarity which gain legitimacy and circulate within the city and Indian nation more generally.

What is interesting to ask here is: Why were these two religious actors able to capture the imaginations of city residents in appealing for continued inter-religious peace? Why did they become the subject of positive local attention and media coverage in a way that state and national level politicians did not? What granted them the legitimacy to act as effective peace-brokers in this particular scenario? The physical alignment of the *Mufti* and *Mahant* was reinforced by their common rhetoric or use of "peace talk." Both men appealed to shared imaginaries of "Hindu–Muslim *bhaīachārā* or brotherhood" which they sometimes rooted in the city's cultural and economic environments through reference to vernacular forms of "peace talk" such as *Gangā Jamuni tahzēb/sanskritī* , or the notion of *tānā-bānā* and the idea that relations resembled the interwoven warp and weft of silk fabric. These narratives were simultaneously inspired by and reinforced local notions of cosmopolitanism.

Vernacular expressions of peaceful dynamics form an important part of this story, not least because they lent the *Mufti and Mahant* collective imaginations to appeal to, but they also placed the responsibility, and the possibility, for maintaining peace within the hands of ordinary city residents. The presence of everyday and associational civic engagements in Varanasi was a critical enabling factor. It is within this context that the *Mufti* and *Mahant* had access to and were able to effectively appeal to vernacular imaginaries of peace. Conversely, governmental actors on the whole were not regarded as legitimate "peacekeepers." Instead, their actions were met with skepticism and they were seen as detached from and unsympathetic to the real concerns of Varanasi life, whether during periods of tension or not.

The strength of the community at the Sankat Mochan temple was undoubtedly also a factor that enabled the messages for peace from the *Mahant* to be effectively received and acted on. This temple is open to people of all castes, classes, religions and nationalities, and its community extends beyond the immediate devotees to the sweet-makers, milkmen, *chāy* wallahs (tea vendors) and flower-sellers who work in and around the temple. A Muslim family supplies the flowers to the temple, and Sikh and Hindu families lead a special procession for the birthday celebrations of Lord Hanuman. Every April a music festival is held in the temple complex. It attracts famous musicians from across India, and is open and free to everyone. The temple has subsequently earned a citywide reputation as an inclusive space. This perception shaped the psychological impact of the terrorist attack on the temple; it was not just regarded as an attack on Hindus, but also the wider community of Varanasi. Similarly, the Mahant's actions were recognized beyond the immediate temple community as respondents from different religions, classes and castes commented on his exemplary performance in the aftermath of the terrorist attacks, both weeks and years later.

It is also worth noting how local experiences of violence and socioeconomic conditions informed the *Mufti's* actions during this time. Previous
incidents of violence and riots in Varanasi have typically hit the Muslim
community hardest. And, given the apparent economic decline in the silk
handloom business, Muslim weavers and businessmen were particularly
eager to ensure that violence did not result. Without undermining what was
undoubtedly an expression of sympathy for the victims of the blasts, actions
by the *Mufti* should also be interpreted as a conscientious strategy to protect
Varanasi's Muslims, by demonstrating his community's solidarity and commitment to the nationalist project and appealing to the wider Hindu
community that everyday peace could be maintained. Given this, I argue
that we need to be sensitive to the way that actions and rhetoric around
peace necessarily concealed patterns of inequality and relative vulnerability
between Muslim and Hindu communities. These local practices of peace
also provoke further questions about the politics of peace and the burden of
responsibility around peace. Varanasi's Muslims arguably had more to lose
and felt an urgency to engage in activities that publicly and collectively
denounced the acts of terrorism in order to preserve local intercommunity
relations (see Heitmeyer 2009). I explore these questions further in Chapter
Six, but suggest here that such actions represented forms of resilience, and
also a degree of acceptance of the unequal conditions, rather than resistance
towards the uneven politics that characterize their experiences of Indian
citizenship.

Finally, the local and national broadcast and print media played a critical
role in lending legitimacy to the actions and rhetoric of the *Mahant* and *Mufti*
as they disseminated the call for calm in the city. The tenor of the media in
general, whilst critical of the terrorists, was particularly upbeat, and portrayed Varanasi as a resilient and united city. These findings are significant in
the Indian context, especially where the newsprint media has been accused
of adopting a communally divisive line (Olsen 2005) and constructing a particularly Hindu nationalist imaginary of society (Ståhlberg 2004). During
India's "newspaper revolution" in the 1980s and 1990s (Jeffrey 2000) the
Hindi language newspapers, regarded as "close to the people" (Ståhlberg
2004, p. 5), have been widely accused of inflaming communal tensions
(Engineer 1991; Hasan 1998, pp. 214–15; Nandy 1995, pp. 33–7; Rajagopal
2001, pp. 151–211), especially the north Indian publications *Dainik Jagran*
and *Amar Ujala* (Rawat 2003). However, after the terrorist attacks in Varanasi,
even the *Dainik Jagran*, which has a local history for being overtly prejudiced
against the city's Muslims (Khan and Mittal 1984; Malik 1996 and see
Froystad 2005 on Kanpur) was notably constructive and balanced in its news
coverage. Along with other daily Hindi and English-language newspapers it

enthusiastically employed the rhetoric of Hindu–Muslim "brotherhood" and *Gangā Jamuni sanskritī* to portray, and in the process contribute towards reproducing, the city's apparently amiable Hindu–Muslim relations.

Newspapers are avidly consumed within Varanasi's homes and tea stalls, but the majority of informants spoke of primarily following the unfolding events by tuning into the 24 hour "breaking news" coverage that was broadcast on satellite television channels such as Siti and Aaj Tak. Along with Pintu and the girls, I spent the immediate hours and days after the bomb blasts sitting on their beds watching events unfold at the temple less than two kilometers away, and at the railway station in the Cantonment area. Transmitted directly into homes, the uncensored images represented the brutal destruction brought about by the bomb blasts. These raw images could have sparked retaliatory fury amongst some sections of society. However, the violence of the attacks was attenuated by interviews with local and national politicians and the commentary of news anchors which almost universally conveyed the point that the attacks by "foreign terrorists" did not present an excuse to disrupt the city's otherwise harmonious social fabric.

In a detailed ethnography, Simon Roberts has examined the impact of satellite television within the spaces of Varanasi's homes and families to show how, since the arrival of the national channel, Doordarshan in 1984 and then satellite television in 1991, television has become increasingly "another member of [the] family" (1999, 2004). The potential of news broadcasting has continued to develop, so that minute-by-minute "on location" reporting of key events has become the norm. I argue that the potential for viewers to witness events as they happened in real time vastly reduced the chance for pernicious rumors to circulate and gain traction in ways that might have inflamed religious differences after the bomb blasts. Key informants often blamed the role of rumors for inciting riots in the early 1990s, when Hindu right leaning newspaper editors published salacious, unsubstantiated stories, for instance about the number of Hindus killed during the demolition of the Babri Masjid in 1991.

In the aftermath of the bomb blasts, newsprint and broadcast media therefore played a central role in enabling the maintenance of peace. The evenhanded if not overtly united tone of news coverage, in conjunction with "live" satellite news reporting, acted to undermine the arguments pushed by some Hindu nationalists that called for hostile and violent reprisals. The neutral timbre of the Hindi-language media in the mid 2000s, compared with its notably biased reporting in the 1990s, is significant. As a mouthpiece of the people and for the people, there was a perceived shift in readers' attitudes towards the consumption of "communal" news stories. Varanasi's residents articulated a sense of collective responsibility in sustaining good intercommunity relations, mindful of the damage that riots and curfews had brought to lives and businesses, and aware of the political motivations that often underscored such events. The deployment and active uptake of a rhetoric based around real and imagined ideas of the harmonious city was central to the reproduction of everyday peace. Yet, as I further argue in the coming chapters, the successful maintenance of everyday peace is often contingent on concealing patterns of inequality and uneven geographies of power.

Conclusion

This chapter has examined a range of responses and events that were articulated in the aftermath of the 2006 bomb blasts in Varanasi, seeking to make peace visible and to begin to answer broader questions concerning what peace looks like, and how peace is socially and spatially constituted. It provided a perspective on different sites, scales and actors through which peace was (re)produced in the city and which are further explored in the later chapters. In particular, it introduced the importance of shared public spaces and local agency, where the legitimacy of local religious leaders was derived through their capacity to authentically connect with the local. And, within an atmosphere of anticipated violence, this chapter demonstrated empirically how peace is a process that is complexly intertwined with both the memory and anticipation of violence.

In this situated geography of peace the actions and strategic partnership of two key actors – the *Mahant* and *Mufti* - were of paramount importance in realizing the continuation of everyday peace in Varanasi. The public relationship of the *Mahant* and *Mufti* profiled across the media represented a powerful statement about peace, because they actively embodied forms of solidarity and trust in place. Common narratives of vernacular peace that were rooted in different scalar realities and imaginations of coexistence bolstered their calls for peace. Conversely, the responses of state actors did not resonate in the local and were not granted comparable legitimacy – city residents actively distanced state actors from local events. These findings also demonstrate the power of "peace talk" across different scales and for different actors in uniting religious communities and preserving everyday cooperation. It shows how narratives of peace are deployed as both description and political strategy. And, in this context, it shows how, by connecting local peace talk with national discourses of secularism around tolerance and coexistence, actors come to represent a form of "practical secularism" (Jeffery and Jeffery 1994, p. 553).

By highlighting the inferior social and economic conditions experienced by the city's Muslims, and the way in which this inequality reinforced the imperative to maintain peace, the chapter reveals something about the politics of peace and how it is constituted through uneven geographies of power. On the face of it, the maintenance of "everyday peace" might appear to denote a condition of blissful intercommunity harmony and tolerance, enabling equal citizenship and justice for all. Yet, as this window on to peace has shown, people are differentially situated and empowered in that process. I suggest that precisely because the city's Muslim minority experienced patterns of increasing economic marginalization and pervasive discrimination within the Indian context, the need to secure ongoing peace proved to be especially vital during this particular period.

I wish to finish with a more positive observation: that the reproduction of peace may also, in some circumstances, represent a generative process, one that may create new possibilities and give rise to new formations. A month after the bomb blasts, Hindu and Muslim tension had once again arisen following a dispute between the two communities over the installation of a Shiva *linga* (religious statue) in a temple adjacent to a mosque in the locality of Hartirath (*The Hindu* April 20, 2006). The local administration called on the *Mahant* and the *Mufti* to assist in mediating the tension. The recognition by the police that these particular religious leaders had the potential to broker the peace is an interesting statement in the aftermath of the terrorist attacks. Indeed, when I talked with the *Mufti* in the Madrasa Yatimkhana Mazharul-Uloom in 2011, five years after the terrorist attacks, it was apparent just how instrumental this period had been in shaping his public profile amongst Varanasi residents from all religious backgrounds, the administration and politicians. Local Muslim opinion may be divided about his religious politics, but all recognized that he now served an important role in representing the "Muslim" community at the citywide level and in protecting their collective interests during moments of perceived tension. The active work to maintain peace in the aftermath of the attacks had therefore given rise to new intercommunity partnerships and lent a new legitimacy to certain personalities. However, whether such partnerships will continue to be cordial and whether such actors will always work towards peace remains to be seen, as agencies can be both constructive and destructive, as I further document in Chapter Eight.

This chapter raised some important questions about how the relationship between society and the state informs the potential to maintain peace, or not, and how unequal patterns of citizenship influentially shape both the imperative and the responsibility to orientate towards peace. I develop these lines of enquiry in the next chapter by exploring the nature of everyday encounters between local Muslims and the state and examining how they negotiate the prosaic political spaces in the neighborhood to articulate forms of citizenship.

Chapter Four
Political life: Lived Secularism and the Possibility of Citizenship

Introduction

The previous chapter documented degrees of distrust and distance between the state and Varanasi's residents and the way that Muslims found themselves on the back foot in an unequal society. This chapter focuses on the prosaic political spaces of Madanpura, a Muslim majority neighborhood, to examine Muslims' encounters with and experiences of the state and society. In doing so, it reflects upon the policy and practice of Indian secularism as it affects the scalar politics of everyday peace, and examines how local Muslims constructed notions of justice, articulated political agency and carved out spaces of citizenship in the everyday. I show that secularism, as a national policy designed to facilitate spaces of tolerance and coexistence, provided a powerful ideal for Muslims seeking justice and inclusion. Contrary to the alarmist discourse of rising Islamic militancy, the vast majority of Muslims in India remain attached to the idea of secularism, and do not seek an alternative ideology. Instead, I document how Muslim individuals and groups take secularism seriously. They have continued to invest in the ideal and have found creative ways to employ the language of state secularism and evoke

Everyday Peace?: Politics, Citizenship and Muslim Lives in India, First Edition. Philippa Williams.
© 2015 John Wiley & Sons, Ltd. Published 2015 by John Wiley & Sons, Ltd.

secular social principles in order to realize the possibility of citizenship and sustain everyday peace.

The first section looks to research that has explored the nature of encounters between the state and historically marginalized groups before turning to the centrality of secularism in structuring Muslim experiences of the Indian state. Secondly, the chapter empirically documents Muslim Ansaris' encounters and experiences of the state in their neighborhood, which contributes to their unequal access to citizenship. This sets the scene for showing, in the third and fourth sections, how, despite partial citizenship and palpable injustices, Muslim residents articulated different strategies towards realizing degrees of citizenship in dialogue with state secularism and the local imperative to sustain everyday peace in the city.

Experiencing the State and the Potential of Secularism

Scholarship examining experiences of the state in postcolonial India charts a complicated trajectory. Initially preoccupied with elite interactions with the state and formal politics (Morris Jones 1966, 1978), attention has increasingly shifted towards understanding the everyday state's intimate and dynamic exchange with society in a broader sense and from ethnographic perspectives that privilege the role of agency and practice (Gupta 1995; Fuller and Harriss 2001; Hansen and Stepputat 2001; Das and Poole 2004). Recently, scholars have paid special attention to the complex and contradictory ways that marginalized caste groups in Uttar Pradesh relate to the state and its architectures of governance, especially democratic governance, in the context of a liberalizing economy. In a political ethnography that uncovers the processes behind the political rise of Yadavs (a Hindu caste cluster) in the town of Mathura and Uttar Pradesh more generally, Michelutti (2008) contends that this caste group has been able to exploit the possibilities of the state and realize political power through the vernacularization of democracy.[1] She shows how, by interpreting and internalizing democracy through local idioms, the political resonates positively in everyday Yadav life and has increasingly come to be perceived as a profitable "community/caste business" as well as an arena in which matters of dignity and honor are debated.

In a study of a different Hindu caste group, Jats, in the district of Meerut, Jeffrey (2010) found that "politics" (*rajniti*) is regarded with cynicism, even though many have seen direct benefits from the state. Jeffrey has shown how, rather than pursuing material improvement through formal political power, Jats actively sought to intensify their political connections via local and informal channels, which offered the highest returns, whilst simultaneously investing in the discourses and practice of education for their children. Despite the fact that they successfully secured the reproduction of class power through channels orthogonal to the state, or perhaps because of this, Jats continued to believe in the state's material and ideological importance (Jeffrey 2010, p. 180). By contrast, Alpa Shah's ethnography of Adivasis (tribal groups) in the state of Jharkhand, shows how one particular tribal group known as the Mundas actively kept the state away (2007). Looked upon with fear and a sense of danger, the state was perceived to embody a form of politics regarded as inferior to local practices of the Munda "sacral polity" (Shah 2007, see also 2010). Together, the three ethnographies referred to here point towards the uneven way in which state politics and policies are interpreted and negotiated by different marginalized communities. Their contrasting ritual and economic status as well as cultural resources, may or may not be easily adapted to "the democratic political game" (Michelutti 2008, p. 219) and may be more or less conducive to negotiating the politics of marginality.

As a framework implemented to minimize and manage religious cultural and numerical differences, secularism represents an important theoretical and practical site around which relations between the state and its religious minorities have become articulated. The apparent "crisis of secularism" in recent decades, and its failure as a liberal ideal to maintain the potential for intercommunity equality and peace, has provoked a series of critiques by political philosophers and theorists (see Bhargava 1998). The "liberal left" position defends secularism and argues that religion and politics belong in different realms. The rise of the Hindu right in the 1980s was regarded as a failure of the secular state and of modernization to keep religion and politics distinct. The consensus is that if adhered to as the practice of tolerance between communities of difference, secularism would offer the possibility of progress, liberty and rationality (Sen 1998, 2007; Engineer 2003). Meanwhile, the more extreme responses reject secularism outright, perceiving it to be a modernist Western incursion into Indian society that does not take into account the immense value of religion in everyday Indian life (Nandy 1998; Madan 1987).

Running through the spectrum of critiques is the prominence afforded to *tolerance* between religious groups as being critical to India's survival and central to the principle of secularism. But, tolerance as "an attitude that is intermediate between whole hearted acceptance and unrestrained opposition" (Scanlon 1998, p. 54) is complex, broadly interpreted in the Constitution and experienced in everyday life. The failure of secularism and the breakdown of tolerance between religious communities was tragically illustrated by the 2002 riots in Gujarat, which involved disproportionate and often state

sponsored violence against Muslims (Varadarajan 2002). Some have maintained that for secularism "to work" it has to function within all sections of
society and not just within the bourgeois sphere of civil society (Menon 2007;
Chatterjee 2004). Others like Nandy dismiss the potential of secularism
outright, focusing instead on the dynamics of individual encounters and propose that the future of coexistence lies with principles of hospitality and "old
fashioned neighbourliness" (Nandy 2007, p. 114).

Notwithstanding the range of critiques that the idea of secularism elicits in
the Indian context, it is possible to see that it underpins the project of
citizenship; the bundle of practices that constitute encounters between
citizens and the state. The responses outlined above contend in different ways
that state secularism is flawed and has failed to fully deliver the conditions
through which India's Muslims might realize equal citizenship. However, these
critiques fail to consider how secularism might actually be lived and positively negotiated in everyday life. More specifically, they neglect to take into
account how the rhetoric and reality of state secularism is interpreted from a
Muslim minority perspective and how these experiences inform their
citizenship strategies. Paying attention to experiences of, and engagements
with, the state and secularism, the findings presented in this chapter touch on
questions of meaning, agency and justice (see Lister 1997; Werbner and
Yuval-Davis 1999). Following McNay, the notion of experience is located
within the social and must be understood "in relational terms rather than in
an ontological sense as the absolute ground of social being" (2005, p. 175).
Before describing some of the other ways in which Muslims engaged with the
state and articulated different kinds of agency and practices of citizenship,
I examine how the state was experienced and interpreted in the local context.

There is abundant evidence to suggest that India's Muslims are more likely
to experience inferior-quality state services and provisions than other socio-
religious groups, in this section I examine how such patterns of unequal
citizenship were perceived and constructed by residents in Madanpura. At a
time of economic downturn in the silk industry, Muslim *Ansāris* from all
economic backgrounds placed the blame on the government, perceiving its
failure to offer robust support and economic protection as a form of neglect.
The government's inaction reinforced wider feelings that the state had consistently failed to appreciate the value of the weaving community, as one
former weaver Bishr Akhtar made clear: "Sari weaving is the art of the hand…
yet the weaver has never been valued for his art." Despite a widespread sensitivity to the government's history of absence within their community and
profession, there were expectations that the state should and could alleviate
the conditions underlying the industry downturn. As Bishr remarked, many
Ansāris consistently placed primary culpability for their destitute condition

on the government, which they understood within a broader spectrum of neglect for their welfare.

> In all areas of eastern UP [Uttar Pradesh] around 100,000 have quit the [weaving] profession because of the lack of demand. The price of the raw materials has gone up, but the product is not valued by consumers. ... So people are becoming bicycle rickshaw drivers, laborers, hawkers and *subzī-wāllās* (vegetable sellers). Many are dying from hunger. There is no support from the government. No education, no healthcare. The government doesn't have any plan to provide for them.

Madanpura residents repeatedly referred to the lack of accessible and appropriate government schools and hospitals, the antiquated and inadequate sewage system, and the intermittent and unpredictable electricity supply. The latter problem is particularly frustrating for handloom weavers whose work requires close attention to detail in reliable electric lighting. Weavers generally reported having just eight hours of electricity a day. The situation was much worse during the summer when demand on the power grid surges as the population of Eastern Uttar Pradesh attempts to stay cool.

Such problems pervaded other *mohallās* in the city and in Uttar Pradesh more generally, which ranks as one of India's least developed states (see World Bank 2002). However, Muslim residents perceived the lack of state provisions in *their* neighborhood as the result of their marginalized position, and notions of injustice were frequently articulated as relational experiences. The perception that Muslims received differential treatment by the state most commonly found expression in relation to majority Hindu experience. In discussions where *Ansāri* informants raised issues about the local and national governments, such sentiments became particularly apparent. This is illustrated in the comments of Nizar Ahmed, a middle-aged Muslim weaver describing the failure of the then Chief Minister and Samajwadi party leader, Mulayam Singh Yadav, to execute his election promises.

> Maybe 1 percent of facilities were given to Muslims and the rest were given to non-Muslims. Mulayam Singh Yadav said that he would open 100 *madrasās*, but he started with 250 Sanskrit Schools and so far no *madrasās* have been opened, he hasn't implemented his promise.

Repeated issues of injustice experienced by Madanpura's residents shaped their perceived social disadvantage, which was often conceived of as worse than that of Dalits (regarded as "untouchables", recognized in the constitution of India as Scheduled Castes and Scheduled Tribes). These narratives point to a key observation underpinning this chapter: residents in Madanpura, whether rich or poor, perceived that they were discriminated against by the government because they were Muslim. It is less relevant whether government provisions were available or not; what is more relevant is the perception of their absence or inaccessibility to the Muslim community. And that this was explained as being the result of discrimination based on their religious identity and minority status.

The popular perception of the state's negligence in weavers' development matters was at odds with the government's narratives. In fact, state discourses showed concern to strengthen the backbone of India's handloom and artisan industries and to implement various measures designed to alleviate weavers' conditions of poverty. Discussions with government officials at the Handloom Development office, Varanasi, and the Department of Industry, New Delhi, in addition to evidence from government documents (Vaghela 2004; Sinha 2005), revealed the government's awareness of the plight of handloom weavers and associated workers. Sarjit Pandey, Deputy Director of Handloom and Textiles in Varanasi, enthusiastically explained the ICICI Health Insurance Scheme and how this was the most recent and exciting government scheme amongst a range of initiatives designed to support the impoverished weaving community (see Srivastava 2009), the majority of whom he acknowledged were Muslim *Ansāris* in Varanasi City. The scheme was designed to financially enable a weaver, his wife and two children to access "the best healthcare facilities in the country" (Sinha 2005). In 2005, the total yearly insurance premium of Rs1,000 was funded by a government subsidy of Rs800 and a contribution by the weaver of Rs200 to provide health insurance with an annual limit of Rs15,000.[2] After August 5, 2006 the scheme operated on the basis of zero contribution by the weaver. When medical treatment is required the weaver can either reclaim his medical expenses incurred in "any hospital" or receive treatment in an "empanelled hospital" where the bills are paid directly by ICICI Lombard through a Third Party Administrator (Sinha 2005).

Throughout my fieldwork in Varanasi, I asked Muslim weavers whether they knew about the HIS and if they had an insurance card. Saleem, a weaver activist in his mid-thirties, was one of the few weavers who had a HIS card. Many people had applied but very few had received their cards. This experience appeared to be shared widely across the city's Muslim weaving communities. In March 2007 the Rajiv-Vichar Manch demonstrated outside an ICICI office in Sigra, Varanasi. They accused the government of "trickery" and protested against the delay in issuing HIS cards to weavers (*Hindustan Times*, March 1, 2007). Weavers were either unaware of the scheme or months later were still waiting to receive their cards. Those with social or political connections seemed more successful in obtaining the cards, but with no "empanelled" hospitals in Madanpura weavers still needed to find the capital upfront to pay for their health care and claim expenses later. The widely held suspicions were that the majority of those benefiting from the government's program were not Muslim weavers but rather the Muslim businessmen from their community who buy and sell silk saris and who were better placed with regards to education and the social connections needed to access the scheme's facilities. As one elderly Muslim and small-scale *girhast* (master weaver) Abdulla Ahmed reported: "Help is there, but rich people from our own community [Muslim *Ansāri*]… make cooperatives and the money never comes to the poor weaver, it gets eaten by the rich people." For many, the state appeared not only absent, but also inaccessible.

Discussions with Aamir Ahmed, an older male Muslim *Ansāri* who was struggling to keep on the three weavers who worked in his small workshop, encapsulated the experiences of independent and master weavers with regard to government initiatives. It was on a return visit to Varanasi in 2008 that I bought up the insurance scheme in our conversation, aware that it was something we had not previously discussed. Aamir was a relatively success-ful weaver now on the verge of retirement, and yet the narrow profit margins he made on each sari, and the practices of delayed or reduced payments, meant that he regarded his family as poor, especially compared to the rich Muslim businessmen who owned sari shops on the main road, and for whom his weavers produced saris. Aamir recalled countless occasions when mem-bers of his own community had come to his door with government related forms and applications. In the beginning he completed them, but the money supposedly promised was never received. His grasp of the details; what government scheme, which government offices and officers and where they were based, was characteristically vague. But, while the state appeared both absent and inaccessible for Aamir in the local context, it was not entirely imperceptible. The state was imaginarily symbolized by Sonia Gandhi and Mulayam Singh Yadav who represented important personalities of the state at that time and whom Aamir held accountable for his family's impoverished condition.

Despite the assertions made by weavers that wealthy Muslim businessmen did know how to access the state and profit, my research shows that even Madanpura's wealthier residents experienced degrees of discrimination by the state. In the simple but smart and relatively large showroom of Eastern Saris, 35-year-old Jamaal, an educated and smartly dressed member of an elite Muslim merchant family, spoke in defense of the *Ansāri* business community. He believed that businessmen had been the target of unreason-ably heavy and unjust criticism in recent years regarding the conditions for weavers. In his opinion, discrimination by the government bureaucracy was not limited to poorer Muslim weavers. In conversation about the alleged illegal import and price-fixing of Chinese silk yarn by Marwari Hindu traders in the city's main silk market, he expressed his disappointment that the government had not acted against the mainly Hindu suppliers of the illegal yarn. Instead, the government had targeted its campaign against buyers, "people like us." Jamaal believed that government actions were deliberately directed against Muslim buyers, while at the same time actively steering clear of any changes in the system that would adversely affect wealthy Hindu traders, who exerted not only significant economic, but also political muscle in Varanasi. "We have complained on many occasions against these traders, but they haven't acted. The government wanted to be involved but it didn't have any plans as to how it could be effective on this issue. For the last 20 years we have been fighting over this issue, but nothing has been happening."

Unlike many poor weavers, Jamaal and his family *did* have access to the state though certain channels. As a cooperative owner he was able to organize meetings and hold conferences through the cooperative system to inform government officials about issues such as price-fixing. Jamaal had connections with the officials at the handloom offices and knew of their locations in Varanasi and Kanpur and at the Ministry of Textiles in New Delhi. He was also able to build on his father's connections and reputation as a member of an old business family in the city. Despite his links and apparently more tangible knowledge of the state, Jamaal still experienced difficulties accessing government representatives on development matters. He recalled how members of his cooperative had travelled to Delhi to meet government officials to discuss price-fixing. They waited all day and were eventually told that a meeting would not take place and that they should leave a memorandum instead. To Jamaal's mind, this episode epitomized the government's persistent neglect of the Muslim community's needs.

In conversations concerning the government's apparent inaction, the condition of India's weavers was often compared to that of farmers. Also regarded as marginalized, the perception was that, as an occupational group, farmers received considerably more support from the government. According to Jamaal, "if there is any interest in the weavers it's zero percent, but for the farmers it's 100 percent." The explanations offered for this discrepancy pointed to the predominantly Hindu identity of farmers but also the considerable vote-bank they represented, evidenced by the number of seats as well as the level of influence held in parliament. In Jamaal's view, "[T]he difference between weavers and farmers is that farmers have the political connections. If the farmers are upset then the government is set to lose many-many seats in parliament, but if weavers are upset then only a couple of seats would be lost."

The perception articulated by Muslim *Ansāris* in general was that the government persistently neglected their social and economic needs. Such neglect was interpreted in relative terms, contrasted with the relative advantage of the majority Hindu population, and recognized as the result of discrimination based not only on their religious identity but also their minority status. The weavers' sense of marginality and limited access to the state were also determined by socioeconomic and educational conditions. For the poorest weavers with little education, their marginality was amplified, if not mediated, by their low status within the Muslim community itself. Despite processes of Islamicization since the early twentieth century in Varanasi, which have elevated the importance of piety (Searle-Chatterjee 1994; Gooptu 2001), it was clear from residents' stories that respect and influence were increasingly commanded on the basis of wealth, in conjunction with educational attainment and religious commitment. Whilst poor weavers lacked the capacities to avail themselves of state schemes, and regarded the state with cynicism, they did nonetheless hold the state accountable for their deprived conditions. On the other hand, richer *Ansāri* businessmen were able to utilize government schemes, for example in setting up cooperatives and extending their local connections with other communities. Their relative economic and educational advantage meant that they were able to capitalize on schemes for their personal advantage whilst claiming to mediate between the state and poorer weavers. Despite these advantages, rich Muslim traders still felt frustrated by the lack of influence they were able to wield at the local state level and perceived the state as distant and discriminatory. It is against this backdrop of unequal access to the state that I turn to illustrations of the ways in which local Muslims continued to engage with the state to make claims to their citizenship. I argue that the potential to realize citizenship depended on not only their vertical relations with the state, but also the quality of horizontal relations within society.

Engaging with the State, Practicing Citizenship

It is argued that subordinated religious groups often lack equal access to the material means of participation (Fraser 1990, Chandhoke 2005). Whilst this was evidently the case for many of Madanpura's Muslim residents, this section builds on the argument that marginalization from the dominant public sphere does not entirely deny the possibility of alternative emergent and marginalized publics (Eley 1994, p. 308). Moving away from state defined, top-down conceptions of citizenship (see famously Marshall 1950) and their inherently patriarchal, Western conception (Lister 1997, Kabeer

2007), scholars have attempted to reformulate our conventional conceptions of citizenship.

It is increasingly recognized that even whilst citizenship speaks to encounters between society and the state, the possiblity of enacting citizenship is fundamentally contingent on experiences of, and positions within the "social" (Isin 2008).[3] Conceptualizing citizenship as produced in and through the social draws attention to the limits of the state in delivering rights, where social groups and relations also mediate experience. It also highlights the ways in which dominant constructions of citizenship may be contested and overturned by political actors seeking justice. Citizenship is therefore conceptualized as an incomplete process; a form of self- and subject-making that is always in the process of becoming, and where those excluded may come to define the terms of their exclusion.

Given the partial and increasingly diminished role of the state, especially in the context of liberalization, other scholars have forced open the concept of citizenship in order to comprehend the plural and shifting ways of being and becoming a citizen (Ong 1999; Lazar 2008). But, expanding the notion of citizenship in ways that may challenge the primacy of the state does not leave the state behind altogether. Subramanian (2003) shows how the parameters of citizenship may be recast in ways that allow citizens to challenge the state's own criteria. She documents the actions of artisanal Catholic fishermen in south India as they sought state recognition "on their terms" rather than through the channels prescribed by the secular developmental state. Similarly, the way that citizens navigate and reconfigure the meaning of citizenship with respect to the state has formed the subject of research in Israel. Stadler et al. document how the practices of religious "fundamentalists" should not be conceptualized as outside or counter to the spaces of national citizenship. Rather, they show how fundamentalists blend and confront religious ideals such as "contribution," "participation," and "sacrifice" with similar ideas promoted by Israeli republican citizenship discourse to create a space for themselves (2008, p. 218).

In the Indian context, Shani (2010) has argued that India's Muslims have strategically negotiated multiple citizenship discourses in order to cultivate a share in the nation since Independence. Her contention is that republican, ethnonationalist, liberal and non-statist citizenship discourses have, over time, presented different opportunities and barriers to Muslims attaining full membership of the nation. As the dominant citizenship paradigm shifted to exclude them in new and different ways, Muslims sought the strategic alternative for best realizing their citizenship. Shani's thesis is important for highlighting the plural and shifting frameworks within which people may position themselves as citizens through consent and dissent in order to realize meaningful participation. However, it does not go far enough in considering how ordinary citizens engage with and negotiate citizenship discourses in their everyday lives, perhaps variously interpreting a particular discourse or simultaneously positioning themselves with respect to contrasting citizenship paradigms. Furthermore, the discourse of secular citizenship is curiously absent in Shani's work.

The following sections build on her insights by seeking to understand how ordinary Muslims in an everyday Indian city negotiate and employ secular citizenship discourses in order to make their citizenship claims. It thereby challenges the idea that citizenship is defined by exclusion, instead shifting attention to the struggles, sites and spaces through which efforts towards inclusion were expressed and negotiated.

Partial Citizenship, Imagining Citizenship as Connection

Muslim *Ansāris*' expectations of the state to deliver basic provisions equally within society, despite a widespread recognition of its failure to actually do so, reflects a sustained faith in the *idea* of the secular state and their taken-for-granted status as Indian citizens. Residents did not alienate themselves from the state, nor did they respond in a form of passive acceptance. Experiences of an inadequate or absent state in critical areas of development in Madanpura, notably in education and health, have encouraged community initiatives to compensate for this service deficit. Like other Muslim majority locales in north India, Madanpura has a precedent of community provision in efforts towards strengthening *Ansāri* identity in the context of Hindu resurgence in the early twentieth century (Gooptu 2001). The developments seen here should therefore be understood as part of a broader trajectory of Muslim self-provisioning, as well as a particular response to the postcolonial state's perceived failure to provide for and protect the interests of Madanpura's Muslims. Since the 1970s, three private hospitals have been established, managed and funded by *Ansāri* businessman and weavers in Madanpura. These are the only hospitals within the *mohallā*. Whilst there is a government hospital in Bhelupura to the south of the neighborhood and another private hospital run by a Hindu organization, the Ram Krishna Mission, to the north, neither of these offer maternity facilities.

The Banaras Public Welfare Hospital and Janta Seva Hospital were established during the 1970s; a period which saw growing economic wealth amongst the Muslim business community, coupled with periods of tension between Hindu and Muslim communities following City riots in 1972, 1977 and 1978.[4] Kalim Ahmad Mahto General Secretary of Banaras Welfare

Hospital and Muslim sari businessmen explained that after the riot in June 1972 he was playing carrom with his weaver and trader friends when they had the idea to open a hospital in Madanpura:

> So we decided then and there to get a receipt [book] and collect donations from people, just Rs.10 from everyone, and more from those who could afford it, until we collected Rs.42,000. We also raised big donations, so that's how we were able to buy this building for Rs.150,000. At the time there were 40 tenants living here so we had to free the house of them and start the hospital. It was registered as a hospital in 1977.

Discussions with hospital founders revealed that motivations behind establishing community health facilities were situated within a climate of growing insecurity in the city. It was decided that a local hospital was needed *within* their own "Muslim" neighborhood where women could give birth safely and those injured in riots could easily obtain medical attention during communally tense periods when, it was perceived, the protection of Muslims in other non-Muslim areas could not be guaranteed. The local commitment to improving the quality and nature of health provision in the neighborhood was underlined by informants' sense of responsibility that as successful Muslim businessmen they should contribute to the social welfare of less fortunate sections of their community. As well as receiving donations from wealthy sari businessmen, the hospitals were financially supported by a cross-section of Varanasi's Muslims as Mohd Naseem, General Secretary of Janata Seva Hospital explained: "Not only the *ghadidars* (sari traders), but also the middle-class people donated money to the hospital. On Bakr eid the skin from the goats is donated to the hospital, which we can sell and raise funds for the poor patients. So poor people also donate..."[5] Both hospitals began as informal and rather basic initiatives, but have since expanded and developed into much more sophisticated ventures which have earned the welcome recognition of the state government.

It is important to appreciate that the medical provision and pharmaceutical facilities offered by the hospitals established through this initiative extended beyond the immediate Muslim community. Patients largely comprised the city's lower socio-economic classes: Muslims, Hindus, Christians and Sikhs alike. Kalim estimated that at Banaras Public Hospital, 60 percent of patients were Muslim and 40 percent were Hindu, all from poorer families. The workforce is similarly secular, with medical and support staff representing different communities across the city; indeed, many of the city's top practitioners, the majority of whom are Hindu and now based at the Banaras Hindu University Hospital, spent a significant amount of their early careers at Banaras Welfare

Hospital. Associates of these hospitals were eager to express their pride in the fact that their institutions, funded solely by Muslims, served a much wider constituency and therefore constituted complementary rather than private welfare spaces. For board members it was critical that their institutions be recognized as inclusive, secular sites for Varanasi's poorer citizens, and not as reserved for Muslims. A member of the Janta Seva management committee reiterated this stance when he later informed me that in 2007, the hospital board had, for the first time, placed a banner on the main road "welcoming" the Kali Puja procession. This procession is taken through Madanpura every year by Bengali Hindus and is characteristically marked by tension between the local Muslim and Hindu communities. When I asked him why this was, he responded: "Because we wanted to demonstrate the strength of Hindu–Muslim unity and by doing so to further improve our relations."

Despite their compromised citizenship, residents did not give up on the project of citizenship entirely. Instead they sought to formulate alternative spaces for realizing citizenship. Not only did this community establish welfare institutions that were complementary to the state provisions, but they were also conceptualized as inclusive, secular spaces, thereby recasting their own experiences of citizenship and restoring notions of tolerance and peace. Within these institutions the management committees keenly recognized their potential to promote local practices of tolerance and coexistence, which engendered new possibilities for citizenship as connection and inclusion. The keen construction of such discourses around cultural connection and hospitality at once represented an evident reality in the neighborhood. But it also reflected a widely held perception by Muslim residents about how their autonomous yet complementary institutions designed to compensate for a *lack* of substantive citizenship might be viably forged and positively recognized by the state *and* society.

"Muslim" Rights to Urban Space?

Societal forces and social pressures were critical in shaping access to the provisions that the state did provide, and they informed Muslim rights to the city more generally. For Lefebvre (1996) "the right to the city" represents not only rights to urban services, such as housing, work and education, but also the right to participate in making "the urban," to both inhabit and transform the urban landscape in one's own image. Struggles over urban citizenship concerning matters of identity, belonging and rights to the city actually involve statements about becoming a producer of urban space and of citizenship itself. Upon learning from a friend about the presence of a relatively new mosque in Madanpura that had remained shut virtually since the day it was opened I was curious to find out more. I arranged to meet the owner of the mosque, a doctor in his early sixties, in his surgery on Madanpura's main high street. During the course of this and subsequent meetings with Dr. Massoud and residents of Madanpura, it became apparent that what had been conceived of as a religious and personal act of remembrance had been transformed into a public act that had come to define neighborhood-wide imaginaries about their right to urban space and the limits to citizenship.

Dr. Massoud's wife died in the 1970s and, in recognition of her final wishes, he dedicated his efforts to establish a mosque in the lane just behind his surgery. For this project the doctor donated one of his family properties for religious and charitable use, and proceeded to have its *waqf* status[6] officially approved. Once formally recognized by Muslim law he proceeded to convert the second and third floors into a mosque, with the intention of opening a medical clinic on the ground floor. In 1999 the mosque was opened and prayer (*namaaz*) was conducted for fourteen days. On the fifteenth day, local Hindu residents physically objected to the presence of the mosque. The police were forced to intervene in the situation and ultimately called a halt to the mosque's activities.

A petition against the mosque was filed at the High Court. This was signed by 145 members of the Hindu community, including local and wider city residents, many of which were local BJP and Rashtriya Swayamsevak Sangh (RSS) party workers and sympathizers. However the High Court dismissed the petition and subsequently gave permission for *namaaz* to be performed as of 20 April 1999.[7] The Court also issued the statement that peace should be maintained and that anyone causing disturbances during prayer times would be prevented from doing so. In spite of this official endorsement, less than two months later, police once again forced the closure of the mosque following complaints from Hindu citizens and recommended the demolition of the religious structure.[8]

The case to reopen the mosque has been pending with the Supreme Court since 2000. When I met with Dr Massoud in 2007 he was still hopeful that religious observance could resume at the mosque but he was exasperated with what he perceived as the government's biased treatment against Muslims. "If the matter was that of a temple then they [the court] would have given the judgment in the middle of the night and made special arrangements."[9] Moreover, for Muslim informants, the public rejection of the mosque by the local community was particularly difficult to digest in a city where the landscape was already characterized by thousands of Hindu temples and the frequent construction of new temples was never publicly impeded by

Muslims. The majority community's refusal to recognize the legitimacy of the mosque meant that a personal religious gesture was transformed into a public debate over the limits of Muslim secular citizenship and their rights to representation in the urban landscape. The sense of injustice experienced by the doctor and neighborhood residents was made all the more acute given the High court's approval for the building.

The doctor's contested mosque inspired neighborhood wide imaginations about Muslim rights in Varanasi. The incident was frequently evoked, often independent of my questioning, to elucidate issues of local tension and injustice between Muslims and Hindus. As one middle-aged Muslim weaver bemoaned:

> [W]e can't build a *masjid* there…[it's] believed that this could create a Hindu–Muslim riot… If you go around the city you will see many temples built on the road which have encroached upon the road and narrowed the passage. And these were built without permission. When we have permission and all the correct articles and the high court order it's still not possible for us to build a mosque.

Forced to confront societal opposition and the local atmosphere of hostility towards his religious act Dr Massoud took care to postscript our numerous discussions about events by asserting his commitment to the Indian nation, and his role as a good citizen. Such a conscientious affirmation of his national identity revealed his sensitivity towards a widespread prejudice that acts against the presumption that Muslims are loyal or good Indian citizens.

> First of all you have to keep in mind that I love my country, I'm a nationalist. I have the same right as other communities, like Sikh, Christian and Hindus. Being an educated person I'm not going to take any kind of action which will affect the nation. We live here, eat here, we have every right to buy houses.

The desire to reproduce religious identity and meaning in the urban landscape was couched within a wider public commitment to the nation identified through the language of secularism and citizenship. For the doctor being a "good Muslim" and a "good Indian citizen" went hand in hand; however, it was by emphasizing the latter that he sought to realize justice and the public recognition of the former. No doubt limited, the discourse of secularism opened up a space of opportunity within which this minority community could assert their claims to citizenship. The reality of state secularism was imagined in contradictory ways. On the one hand, the doctor perceived the lower levels of the state, particularly the police administration to be prejudiced against the Muslim minority. Precisely because the police force was predominantly composed of Hindu officers and embedded within the local Hindu right social milieu, the doctor did not invest expectations in the administration to act in a secular manner. But, on the other hand, the doctor believed that secular principles would be enacted within the higher levels of the state judiciary and he continued to wait patiently for the court to enact the correct decision, one that he believed would ultimately facilitate the realization of justice in the neighborhood.

Conclusions

This chapter develops an understanding about everyday peace by foregrounding the scalar politics of peace, the important role that policy and practice play in shaping everyday peace on the ground and the articulation of local agencies. It has highlighted the discourses and actions through which individuals and groups conceptualized notions of justice and the "Other" and sought to negotiate and actually realize their citizenship by appealing to, rather than rejecting, the idea of secularism. By reflecting upon a range of sites through which Madanpura's Muslims experienced and engaged with the state and society it advances four key ideas that inform the reproduction of everyday peace.

First, it documented how patterns of unequal citizenship were experienced and interpreted by Muslim *Ansāris*. Rich or poor, they perceived and experienced the absence and inaccessibility of the state through a relational lens of being a minority Muslim community. Perceptions of, and encounters with, the state were calibrated in line with education and economic conditions. Poor weavers narrated the state's presence through its material absence, and regarded the government bureaucracy with apprehension and suspicion. However, they continued to hold the central government accountable for their deprived circumstances. Wealthier business classes and social reformers, on the other hand, found the state and its bureaucracy evidently more tangible. They actively engaged with the reality and idea of the state, despite acknowledging its discriminatory tendencies. *Ansāris* sought government recognition for locally established institutions in order to grant them legitimacy as modern civic institutions, within and beyond the neighborhood. However, they regarded politics and the state as constituting a murky world best avoided, one that more often favored majority Hindu concerns and causes.

Secondly, the chapter highlighted the agency of local residents over community or political leaders in shaping everyday peace. Importantly, it showed how patterns of unequal citizenship informed the kinds of agency articulated by local residents as well as their expectations for what everyday peace looks like.

Local residents chose to autonomously establish community welfare institutions within counter-publics rather than publicly contest the state's apparent discrimination. I argue that these actions privileged minimizing Muslim visibility in the city, and not challenging the status quo. As such, they constituted forms of defensive agency, which were informed by degrees of pragmatism, acceptance and resilience to both prevent intercommunity tensions and to secure, as well as improve, the future capacities and ambitions of *mohallā* residents. Defensive agencies were the pragmatic option to address patterns of inequality and sustain everyday peace for two reasons. Firstly, as one of many deprived and marginalized communities in Varanasi as well as in Uttar Pradesh more widely, residents were cognizant of the government's inability to respond to their demands. This was in part due to a lack of political leverage and representation by *Ansāris* within local and national political circles, which contributed to feelings of alienation and of being on the edge of the politics that mattered. Secondly, and more importantly, embedded as they were within the north Indian social and political context, Muslim *Ansāri* practice was constrained by the potential for violent feedback from the Hindu right amidst accusations of minority appeasement. Held up as an ideal, secularism offered the potential for Muslims to realize their rights as equal citizens. However, it was accepted that the politics of violence and Hindu majoritarianism inevitably mediated the degree to which such rights could be actually experienced, and care was taken not to provoke intercommunity tensions. This shows how the struggle for inclusion is carefully negotiated within different settings and across scale.

Thirdly, the chapter contributes insights into the politics of citizenship, and explores reasons why this particular marginalized group apprehends the state as simultaneously discriminatory *and* desirable, and what that means for everyday peace. One answer might lie in the state's role as a last resort for a population that has experienced persistent and pervasive patterns of discrimination within society and marginality within local political networks. On the one hand, the state was imagined as an ideal, as constitutionally secular and democratic and in which Muslims as legitimate Indian citizens did have an equal right to justice and the conditions demanded of "positive peace" in spite of their lived realities. On the other hand, Muslim *Ansāris* did not respond to the state's failure to provide adequate welfare provisions through protest and outright public disaffection, because attracting public attention would have entailed the risk of further social discrimination and the potential for intercommunity tension.

Whilst *Ansāris* believed that in theory the state acted in the interests of justice and equality, in practice, without political representation or formal channels such as the education and employment reservations granted to lower-caste categories, the state's (and society's) positive recognition could not be taken for granted. Residents accepted that realizing equal justice was elusive, yet actively protected everyday peace. These findings challenge the binary notion of "negative" and "positive" peace, where the former denotes the absence of violence and the latter refers to *both* the absence of personal violence and presence of social justice (Galtung 1969). Looking at the conditions which shape everyday peace on the ground in Varanasi forces us to recognize that peace as a lived process is actually constituted through struggles for justice, citizenship and inclusion. However, as I go on to argue in subsequent chapters, these uneven patterns of justice and citizenship inform a precarious form of peace that has its limits (Chapter Seven) and is open to manipulation (Chapter Five).

Throughout the chapter I have emphasized that citizenship claims actively drew upon the language of secularism whilst simultaneously challenging the entrenched notion of victimhood that limits the idea of Muslim agency. Despite its drawbacks, secularism did provide opportunities for Madanpura's Muslims to forge alternative spaces for meaningful participation in the nation and everyday city life. Reminiscent of Nandy's (2007) approach to Indian secularism, Muslim *Ansāris* also asserted their citizenship through narratives and practices of connection, shared national identity and hospitality that proved hard for the Hindu majority to reject and misrepresent. Such actions reinforce the claim made by Isin (2008) that it is only through becoming social that civic and political rights can be made possible and matters of injustice exposed. But, it also diverges from Isin's central thesis that citizenship is defined through exclusion. For Varanasi's Muslims, citizenship in practice actually concerns the ongoing quest for inclusion, and in this setting the possibility of citizenship is bound up with the reproduction of everyday peace. Finally, the chapter has started to develop an understanding of multiple and plural Muslim citizenship strategies that are enacted simultaneously across different spaces and scales within the state and society, not only through time (cf. Shani 2010), where notions of the self and citizenship are always in the process of becoming. In the next chapter I turn to the civic spaces of the neighborhood to explore encounters between Muslim residents and Bengali Hindu neighbors and what this means for the balance between peace and violence and the scalar politics of agency in reproducing everyday peace.

Notes

1   A similar argument could be made for Dalit Chamars documented by Ciotti (2010) in eastern Uttar Pradesh.
2   In 2007 a weaver might expect to earn between Rs 80 and 150 per day. On April 1, 2007 Rs100 was approximately equivalent to £1.18.

3   Naila Kabeer (2007) points to a similar shift in focus in her discussion about the importance of "horizontal" relationships within society as well as "vertical" relations with the state.

4   In 1972 Muslims were taking out a procession protesting against Aligarh Muslim University Bill. The police had invoked section 44 and hence tried to stop the procession, resulting in the eruption of communal violence (Engineer 1995, p. 199). The riot of 1977 occurred after a dispute arose when Bengalis carrying the Durga Puja procession (sponsored by Golden Sporting Club) through Madanpura allegedly attempted to take a new route through the centre of Madanpura. The violence and curfews which followed lasted for two weeks (Khan and Mittal 1984; Engineer 1995, p. 199).

5   Bakr-Eid is an Islamic festival to celebrate the willingness of Ibrahim to sacrifice his son as an act of commitment to God. It is a time when family and friends meet and share food, and visit the mosque to give thanks to Allah.

6   Waqf is a permanent dedication by a Muslim of any moveable or immovable property for any purpose recognized by Muslim Law as pious, religious or charitable.

7   Court Proceedings; Dr. Massoud's collection April 20, 1999.

8   Kazi–e-Sahar et al. Letter correspondence to Vice President Vikas Pratdhikran Varanasi June 4, 1999; Dr. Massoud Interview March 1, 2007.

Chapter Five
Civic Space: Playing with Peace and Security/Insecurity

The Procession

It was around sunset during *Ramzān* when Pinku and I arrived in the neighborhood of Madanpura in Varanasi. Men and boys could be seen inside sari shops breaking their fast with *īfta* as we moved through the narrow lanes to reach the Golden Sporting Club (GSC), a predominantly Bengali Hindu-run organization. Located just beyond the *mohallā* boundary, the site of the club where the procession would emerge was conspicuous by the intense police presence in the lanes. These policemen appeared not just in their everyday attire, but were also equipped with various forms of body armor; baseball helmets, riot hats, even motorcycle helmets complemented by a similarly creative collection of *lāthīs* (sticks) and baseball bats. It was still early evening and most groups were at ease on the platforms of houses waiting for work to begin. Other more senior policeman consulted pieces of paper or spoke with members of the Civil Defense Committee, identifiable in their white baseball caps. Inside the *pandāl* (temporary enclosure) which housed the *mūrti* (statue) the atmosphere was electric and full of anticipation. At the center of the crowd a senior police officer endeavored to command the space and the attention

Everyday Peace?: Politics, Citizenship and Muslim Lives in India, First Edition. Philippa Williams.
© 2015 John Wiley & Sons, Ltd. Published 2015 by John Wiley & Sons, Ltd.

of club members. One policeman who stood close to me asked a member of the Puja club what time the procession would start, to which a particularly vague and non committal response was uttered: "Perhaps around seven, we'll have to see." It was clear that the GSC members were the ones in charge here.

The procession actually got underway just before eight o'clock in the evening. Ten drummers led the path of the Goddess Durga; the enthusiastic rhythm of their beating was matched by the ferocity of chanting and energetic dancing by the procession members that followed close to the idol. Members of the Puja club had described their feelings of pride (*dum*) and courage (*ulās*) when participating in the event. Huge generators supported the voltage for spotlights which powerfully illuminated the route. Riding behind the statue were two sizeable speakers and the club president with microphone in hand welcoming and thanking onlookers. The sights and sounds of a usually quiet lane had been utterly transformed. Initially the procession passed the homes of Bengalis where *prasād* (sweets) was offered to Goddess Durga and red petals were showered from above by women and children who were perched on crowded balconies. As it crossed into Madanpura, now passing predominantly Muslim homes, the audience was much less receptive and interactive, but nevertheless still curious. Muslim men stood in twos or threes around front doors, or congregated at junctions in the lanes to witness the procession pass. Here bamboo fences cordoned off the procession route and anyone contravening the blockade was quickly reprimanded by Madanpura's more elderly inhabitants. Policemen lined the lanes along the route, they surveyed from the rooftops and balconies and they physically escorted the procession through the narrow lanes in force.

The volume and intensity of the procession reached a crescendo as the idol emerged at the main high street and members of the GSC spilled out on to the road, now dancing more enthusiastically than ever. Huge speakers erected on the side of the road blasted out upbeat music. The unwieldy statue was paraded to the center of the *mohallā* where it was spun in large staccato pirouettes to face the crowd gathered around the Goddess Durga. The crowd circled the statue in layers; anxious-looking policemen patrolled close to the statue, teenage Hindu boys and men previously forming the procession and still dancing now joined hands in a large circle behind the police; ringed behind them were members of the Civil Defense Committee; next, police stood in layers, two or three deep, sometimes interrupted by the odd spectator (see Figure 5.1). Towards the back, Muslim men milled around chatting, drinking tea and occasionally glancing towards the activity: the more laissez-faire policemen joined them, or stood in their own groups. At the borders of Madanpura, blockades had been erected by the police. Apparently parked at the ready, but no doubt mainly designed to create an environment of

Figure 5.1   A member of the Civil Defense looks on as the Goddess Durga is paraded in the main high street. Source: Author (2006)

intimidation were three police jeeps, two large police vans, a fire engine and a lorry belonging to the Rapid Action Force (RAF), now emptied of its troops who were patrolling the streets. By nine o'clock the procession finally moved on its way towards Godaulia. As the crowd moved out of Madanpura into the Hindu-majority neighborhood of Janganbari, the fervor and tempo of the procession waned considerably as it left the Muslim neighborhood behind.

This chapter focuses on the civic spaces of Madanpura to show how everyday peace is actively compromised by the actions of local Bengali Hindu residents, for whom playing with peace affords a certain degree of power within the neighborhood and wider city politics. The procession outlined here, of the Hindu Goddess Durga by the Golden Sporting Club (GSC) through Madanpura is one of two annual events in which Bengali Hindus paraded their religious idols through the Muslim neighborhood en route to the River Ganges. Just a couple of weeks later another Bengali Hindu Puja club, the Nau Sangha Samiti (hereafter Nau Sangha), paraded their image of the Goddess Kali. In a similar fashion to that described here, they dominated the lanes and main high street of Madanpura and attracted an overwhelming police presence for the evening. Both processions contain capricious histories in the *mohallā*, having been at the center of tension and occasional violence between local Hindus and Muslims in the past.

These processions annually attract the attention of the state government, which deploys hundreds of police within Madanpura in the production and reproduction of everyday peace in the *mohallā* and the city more widely. Understanding how these particular religious rituals interrupted and threatened to disrupt everyday peace in the *mohallā* prompts us to consider the relationship between space as socially constituted, and peace as the precarious performance of social relations. It reveals how spaces of peace are vulnerable to manipulation by certain actors and for whom playing with peace can engender local influence and authority. Given their perceived threat to peace, these processions took on much wider significance within the city, which served to temporarily sharpen notions of religious differences and foreground the active need to maintain citywide harmony.

If space may be conceptualized as the recognition of territory through social practice, it follows that a spectrum of social and material practices inevitably configure and transform spatial patterns and dislocations. In this way, certain spaces may become marked by ethnic (Eade 1997; Anderson 1991) or caste identities (de Neve 2006), which often reflect relative practices of power articulated by minority and majority populations (Yiftachel and Yacob 2002) and which may shift through time. As de Neve and Donner (2006) assert, ethnographic research within various localities can reveal the micropolitics involved in the creation of such relations of power. They argue, drawing on Low and Lawrence-Zuniga (2003), that ethnographic analysis should examine the concept of space as a tactic and/or technique of power and social control (2006, p. 4). Such an approach emphasizes the contingency and fluidity of place, shaped through ongoing struggles; access to resources, the politics of difference and contested social relations are all played out through specific spatial practices.

Focusing on life in Madanpura affords the opportunity to interpret difference as a "located politics of difference" (Jacobs and Fincher 1998, p. 2). This may be useful in emphasizing not only how persistent power structures can unevenly shape urban lives, but also "the ways in which such structures are, in turn, shaped by the contingent circumstances of specific people in specific settings." It also attends to the various ways in which "that specificity – both social and spatial – can transform structures of power and privilege; the ways oppressed groups can, through a politics of identity and a politics of place, reclaim rights, resist, and subvert" (Jacobs and Fincher 1998, p. 2). And, at other times articulate practices of resilience, patience and acceptance, agencies that do not explicitly contest and seek to publicly overturn overt power structures (Katz 2004; Mahmood 2005; Scott 2009).

The remainder of this chapter first describes the recent history of Madanpura to show how local, national and geopolitical events have contributed to the reification of the neighborhood's material and conceptual boundaries along religious lines, whilst not entirely effacing or undermining the potential for everyday intercommunity encounters. In light of this context I return to the Puja clubs to examine how historical episodes of violence between Bengali Hindu and Muslim communities in the course of these processions have remained unresolved. The third section shows how the suspension rather than reconciliation of tensions between these groups functions to endow the Bengali Puja clubs with a certain degree of influence and power over local space, which is actively reproduced by manipulating and playing with notions of everyday peace. The focus of the fourth section shifts towards the mechanisms deployed to maintain peace and shows how the drafting of significant police reserves within the neighborhood was deemed necessary to maintain local security. However, as the chapter shows, the presence of police bodies as peacekeepers temporarily reconstituted local space and notions of insecurity, bringing contradictory outcomes for different groups. In the final section I show how Madanpura's residents privileged local, informal initiatives for peace, before drawing conclusions about the ways in which the precarious reality of peace may be manipulated by and for different groups.

Reconfiguring Spaces of Security: Consolidating the "Muslim *Mohallā*"

The apparent necessity for heavy police protection during these Hindu religious processions, as outlined in the opening vignette, appears striking. It is notable that even for Muslim processions such as during Muharram, equivalent levels of state security were not drafted in. The particular histories of the Kali and Durga Pujas in Varanasi are no doubt integral to the answer, but these should be situated within the recent transformation of Indian sociopolitical economy and normative narratives which associate Muslim bodies with notions of threat, danger and insecurity.

The cultural project of *Hindutva* was strategically and ideologically positioned around an anti-Muslim stance, which portrayed the minority Muslim population as a threat to the purity of a Hindu *rāshtra* (nation). Appadurai conceptualizes the "majority" community's anxiety of the "minority" as a "fear of small numbers" (Appadurai 2006, p. 11). He argues that small numbers at the national level can "unsettle big issues where the rights of minorities are directly connected to larger arguments about the role of the state, the limits of religion and the nature of civil rights as matters of legitimate cultural difference" (Appadurai 2006, p. 73).

At the city level this is interpreted through electoral politics, neighborhood boundaries defining "us" and "them", and a localized concept of "Muslim danger" which was increasingly shaped by the geopolitical rhetoric concerning the War on Terror and appropriated by the Indian government (Jones 2008). In a sense the "Muslim *mohallā*" represents an urban expression of the challenge to the majority community's realization of "purity" (Appadurai 2006, p. 11), and by implication also local and relative power in the neighborhood. Appadurai stresses that the "neighborhood" represents a conscious construction which is historically and therefore contextually grounded. It results from the assertion of socially organized power acting over places and settings, perceived as potentially chaotic or rebellious (Appadurai 1996, pp. 183–4). It also worth noting that in India's villages, small and large towns alike, there is often talk of "Bengali," or "Muslim," or "Bihari localities" marked out for their difference, but rarely reference to the "Hindu neighborhood." As the collective majority "Hindu" forms the reference point, "the norm," while minority groups suppose a threat that needs containing. Just like Muslim individuals become loci of "Othering," so representations of the "Muslim *mohallā*" reflect particular power hierarchies between majority and minority communities in India.

The notion of difference in relation to perceptions of security, extensively characterized discussions around the social landscape of Madanpura; its composition had altered considerably over the last 40 years and markedly in the last 20. Since the decades immediately after the Second World War a considerable population of Bengalis came to occupy houses near to the *ghāts*, particularly in the *mohallās* of Deonathpura, Sonarpura and Bhelupura adjacent to Madanpura (see also Medhasananda 2002, pp. 732–68). As the city of *moksha* (release from the cycle of rebirth), Varanasi proved an attractive location for retirees from Kolkata (formerly Calcutta). A large proportion of the community that settled were educated professional families who supported and founded clubs for cultural activities, primarily Durga, Saraswati and Kali Pujas (Kumar 1988, p. 217), which continue in the city today.[1] As in many north Indian cities Bengalis held a reputation in Varanasi, and particularly amongst Madanpura's residents, for their prominent cultural pursuits and their educated and ambitious children, who typically left Banaras to pursue higher education and enter professional occupations. But, as many residents looked back to Kolkata and elsewhere for the opportunities that Varanasi was lacking, many Bengalis seemed anxious about their declining status in the area. This has been compounded in recent decades by the upward social mobility of the Yadav caste group in the area who progressively purchased properties and asserted an increasing dominance in the lanes south of Madanpura.

Since the 1970s the Assessment register at the Municipal Corporation records have been illustrating a steady decline in the proportion of Bengali-owned properties and a relative increase in Muslim landlords across the *mohallā*. In 2006–2007 households in Madanpura had become almost entirely Muslim-dominated with just 20–40 Hindu families remaining in areas around Maltibagh, Pandey Haveli and Reori Talab. The growing consolidation of Madanpura as an almost exclusively Muslim neighborhood may be seen as the outcome of a rising fear and insecurity experienced by Muslims *and* Hindus of "the Other." The neighborhood was developed in the late 1970s within a polarizing political climate and a dynamic economic environment. This had gradually endowed some with the conditions for mobility in the silk industry, and the associated capital to invest in property.

Conversely, the trend towards out-migration for Bengalis was in response to perceived growing insecurities, stimulated by an increasing frequency of altercations between minority and majority communities in the neighborhood and wider city, as Bengali informants readily reported. This pattern continued throughout the 1980s but accelerated conspicuously between 1991 and 1993 when a series of riots, subsequent curfews and rumors acted to divide communities during the "Ayodhya dispute." Rajesh, a Bengali Hindu in his early forties, and a member of the Golden Sporting Club talked about the fear he had experienced during this period when he lived to the west of the neighborhood.

At that time I was a Hindu chap living alone in the Muslim area of Reori Talab… I was very scared of being in the Muslim area in 1992. At the time when the Babri Masjid was demolished Muslims came to my place with knives and daggers. We knew that something was going to happen so we moved the ladies of the family to a safe place.

The decline in the number of Hindus coincided with the immigration of Muslim families. The increasing currency of Hindu nationalist politics in the 1990s gained strength from mobilizing supporters around an anti-Muslim stance. This chauvinist nationalist rhetoric overlaid and inspired local city dynamics. Muslims often expressed their experiences of discrimination at the hands of police (also see Chapter Six). Such experiences inevitably reinforced growing feelings of insecurity amongst Madanpura's residents as they faced more frequent periods of intercommunity tension. The atmosphere of uncertainty and mistrust concerning their safety was expressed in local urban space. Muslims increasingly concentrated in Muslim-dominated areas, perceiving these to afford greater security during moments of local or national intercommunity discord.

Driving the demand and possibility for immigration was the increasing involvement of some members of the Muslim community in business aspects of the silk sari trade, which had experienced particularly buoyant periods during the 1970s and 1990s. Attracted by the *mohallā's* expanding trade and familial networks, Muslim businessmen were known to offer far in excess of the market price for property. In 2006–2007 Muslim buyers reportedly outbid their Hindu counterparts in the competition for available houses on the edge of the *mohallā*. Hindus living on the outskirts of Madanpura regarded the progressive expansion of the Muslim *mohallā* and encroachment on their territory with great suspicion and apprehension. The speculative and actual changing hands of real estate along the *mohallā* boundaries thereby represented an active site in and through which religious identities were contested and reified.

Pandey ji, a local Hindu shopkeeper whose own business sits near the boundary with Madanpura, described the typical dilemma experienced by Bengali families looking to sell up. Faced with lucrative offers for their property by Muslim buyers they would suffer acute pressure from the Bengali Hindu community condemning their decisions to sell to Muslims, especially given their close proximity to the sacred Hindu River Ganges. He told me about his uncle's house which he sold for Rs 28,000 to a Hindu family in 1981 was now worth Rs 700,000. The house contained a temple of Lord Shiva and a sacred well and lies on the neighborhood boundary. His uncle had wanted to sell the property to a keen Muslim buyer offering a very good price but his close family prevented the sale from happening. Pandey ji noted that at this time there were no Muslims on the right side of the *galī* whereas today many of the properties were home to Muslim families, however in recent years the Nau Sangha had become increasingly interventionist in the area:

> [T]hey do not let them [Muslims] buy houses any more... You see where that shop is? A Bengali had wanted to sell the house to a Muslim, but the Nau Sangha interfered and wouldn't allow it so they took it to the court and it was decided that that the house should be sold to a Bengali and whatever the Muslim buyer had paid should be returned. Right now the tendency of Hindus is not to let them [Muslims] get closer to the river. They don't want them up to the *ghāts*. If you see at present, there is not a single house of the Muslim community all along the *ghāts* from Assi to Rajghat.

The consolidation of Madanpura as a majority Muslim *mohallā* had arguably reinforced the notion of difference between the communities. With reference to the *galī* adjacent to his shop Pandey ji stated how it was universally recognized as "the border between Hindus and Muslims." Pandey ji suggested that the Bengali groups behind the GSC and the Nau Sangha also actively pressurized Bengali Hindus not to sell their properties outside of the community. These local contestations over land and property were inspired by and shaped party politics. This was particularly apparent during the BJP government of the mid to late 1990s when the Nau Sangha was directly aligned with its *Hindutva* ideology and played an overt role in preventing Muslims from purchasing border properties. In the late 2000s the club's procession also proudly displayed a saffron banner depicting the lotus emblem of the BJP, thereby affirming its political orientation.

The increasing uniformity of the *mohallā*'s religious identity should be understood within the context of everyday cosmopolitan encounters and interactions that also shaped its public spaces. On the main Reori Talab–Luxar and Assi–Godaulia arteries which run through the *mohallā*, the intensity of population flows from different ethnic, religious and national groups had progressively increased in recent decades. The "silk" sari businesses that liberally populated the main high streets, as well as the many lanes leading off these, were also expanding and attracted business from buyers across north India as well as smaller numbers of international tourists. Non-Muslim *dudh-wāllās* (milkmen), *makan-wāllās* (butter vendors) and *subzī-wāllās* (grocers) amongst others were also not infrequent sights in the lanes. The augmentation of globalizing processes reinforced the presence of cosmopolitan circuits and also created new spaces within the *mohallā*, where differences were negotiated and traversed.

Everyday encounters and relations evidently persisted between Muslim members of Madanpura and non-Muslim inhabitants of Varanasi through political, social and economic networks (see Chapter Four). Many informants positively referred to their personal friendships with Bengalis residing near Madanpura's boundary. A local Bengali Hindu affectionately known as "Māmā jī" ran a homeopathic clinic in one of the lanes that delineated the border between Madanpura and Pandey Haveli. The vast proportion of his patients were Muslim *Ansāris* from Madanpura. Observing interactions in his clinic during our many visits, it was clear that Māmā jī had a natural rapport with patients and had established good relations with many. On the wall behind the doctor's wooden desk was an image of the Dargah Sharif at Ajmer, a popular Muslim shrine, which was illuminated with multicolored flashing lights. As Māmā jī eagerly recounted his visit to the shrine, it was clear that this pilgrimage memento also communicated to his patients his cosmopolitan outlook.

The doctor also had strong friendships outside of homeopathic practice. Mohammed Sadiq, a respected Muslim member of Madanpura in his late forties and owner of a small *lāt ka māl* (defective/seconds) sari business, regarded Māmā jī to be a particularly old and close family friend. They had known each other since childhood days when they would play in the lanes outside Sadiq's parents' house on the occasions that the doctor came to visit his uncle from their home in Ballia, just outside of Varanasi. In the 1970s when Māmā jī came to acquire part of his uncle's property and moved to the neighborhood, their friendship grew stronger. Both men described how they would "eat together, make food together, and enjoy times together." In fact, it was during the wedding of Sadiq's daughter that I was first introduced to Māmā jī. As a non-familial male and not least a Hindu, he was privileged with access to the ladies' rooms, a point Sadiq made clear when expressing the extent of their mutual respect and friendship.

Stories of friendships between Muslims and Hindus were not uncommon in Madanpura, but neither did they characterize the norm. There often seemed to be more general "talk" about intercommunity friendships than actual realities revealed. Rather, the strength or prevalence of intercommunity bonds was very much contingent on individual personalities and circumstances, which created the possibility and sometimes necessity for intercommunity relations to develop.

Violent Episodes, Suspended Tensions

In processing through the "Muslim *mohallā*" the performance of the Bengali Hindu Puja clubs constituted a determined articulation of power and influence over local space, especially in the context of the shifting social and material boundaries with Muslim *Ansāris*. Numerous other *pujā* processions took place elsewhere in the city on the festival of Dussehra[2] and yet none of these events were surrounded by comparable hype or influence. Importantly, the suspension rather than reconciliation of tensions following previously violent interactions during the processions endowed these clubs with a degree of potential power. This potential was however, contingent on the clubs' members indulging in and continually reproducing their religious processions as "political" events. How religious processions were constructed, performed and received by different parties, at different times, therefore provides a kind of barometer for interpreting local intercommunity relations.

The public cultural life of Varanasi is characterized by religious processions and festivals; it is often quipped that a significant festival occurs every day of the year in the city (Singh 2002). Such processions through urban space and over time are both indicative of the changing nature of social interactions as well as implicated in that change. Freitag (1989a, 1992) has explored how collective identities and structures of power are constructed and contested in Varanasi's public arenas through processions and protests in particular. In the early nineteenth century sources she finds a number of revealing references to collectively observed ceremonies. Whilst these might be fragmentary she contends that "they suggest a world in which the referents, ostensibly religious, nevertheless attracted participation by most of the population, including the 25 per cent composed of Muslim artisans" (Freitag 1992, p. 25). Indeed, the shared civic sense fostered by public ceremonies is suggested by Acting Magistrate Bird when he noted in 1809 that "the religious ceremonies of the Muhammedans and Hindus are so inseparably blended" that any attempt to "disunite them" would constitute a "new arrangement" (Freitag 1992, p. 206).

These reported religious processions of the early nineteenth century may be understood as engendering and reproducing practices of peaceful inter-community sociality which contrast sharply with the picture I present from an early twenty-first century procession, of intercommunity uncertainty and hostility. A brief insight into the histories of the Puja clubs and their previous conflicts is necessary to situate neighborhood events in their contemporary urban and sociopolitical context. The public celebration of Durga and Kali Pujas are both relatively recent additions to Varanasi's festival calendar and date back to the mid-twentieth century. The expatriate Bengali community had sought to recreate the public Pujas of Kolkata which gained popularity in the nationalist period after Independence (Kumar 1988, p. 218).

The processions follow roughly parallel agendas; both were based on the organization of voluntary committees and the donations of the local communities, the entertainment programs involved popular Banaras musical styles featuring *nagāḍā* (drums), and the *pūjā* rituals were characterized by procession and immersion. The Durga Puja procession carried out by the GSC and the Kali Puja procession organized by the Nau Sangha both originated in civic religious organizations located in the predominantly Bengali Hindu areas which bordered Madanpura and were founded in 1970 and 1968 respectively. The notable divergence between these *Pujā*s pertained to the spatiality of the processions, as they began from different locations and wove unique routes through the narrow lanes of Madanpura, before emerging on to the *mohallā*'s main high street. From here, both processions continued towards Godaulia in the heart of the city, before ultimately reaching the banks of the River Ganges at Dashaswamedh *ghāt* where the images were immersed to complete the religious ritual.

In the nascent stages of the GSC it conducted a couple of annual, small-scale processions through Madanpura before it ran into financial difficulties. In 1977, after a two-year break the procession was resumed, but tension surfaced during the proceedings, between some Bengalis and local Muslims concerning the route. It is alleged that at a T-junction in Madanpura the GSC wished to take the image along the *galī* to the left; this more direct route passed a Kali temple and traversed the center of the Muslim neighborhood. A group of Muslims objected to this decision and insisted that the procession continue to the right, a slightly narrower and less direct route which mainly passed Hindu homes (Hildreth 1981). As the dispute erupted the police administration apparently joined their hands (*"hāth jorne lage"*) and begged the Bengali club to proceed to the right and avoid provoking trouble. The GSC defied the administration and the Muslim community's wishes, instead regarding it their prerogative to follow the chosen route through what they conceived of as "public space" that could not be claimed by any one faction.

With the situation reaching stalemate, efforts were made to achieve a compromise and a meeting between the police, community leaders and a local politician ensued. In the meantime, rumors spread that Muslims had attacked the image with stones, and fighting started as a group of Hindus gathered on the main road and began breaking into Muslim shops. That evening the police became involved and a curfew was declared at eight o'clock. The violent aftermath and curfew lasted almost a fortnight; in that time four Hindus and four Muslims were killed. Almost a year later in 1978 the idol was eventually released from police custody and immersed. The period was looked upon with grave regret by Muslim informants; the sari shop owner Mohammed Sadiq regarded it as a *"dāg"* (black mark) on the Muslim community and many others I interacted with were eager that such events should never be repeated. A court case ensued by the GSC and Madanpura residents concerning the route of the procession. The high court granted permission for the GSC to maintain the traditional path through the neighborhood, however the district administration believing this to be an unworkable solution persuaded the GSC to follow an alternative less controversial route, which they have continued to do since the 1980s. However, the need for a clear passage of the idol has continued to provide cause for conflict, as I outline later.

Violence during the course of the Kali Puja was a much more recent phenomenon: it occurred in 1991 against a regional backdrop of increasingly discordant religious politics and the rise of the Hindu right. Local Muslim *Ansāris* and Bengalis both described how during this particular procession the participants had been exceptionally vocal, shouting pro-Hindu slogans and firing crackers as they processed through the lanes. The trouble began when one of these crackers burst in or near a *chāy* shop in Madanpura, much to the disapproval of the shop owner who, along with other Madanpura residents, openly objected to the gesture. The situation between procession members and Madanpura inhabitants became hostile, forcing the procession to be suspended and the image to be temporarily stationed in the lane. Amidst this atmosphere rumors circulated outside the neighborhood that Muslims had thrown stones at the Kali idol and broken it as an act of reprisal for the cracker being fired.

On the basis of these allegations, the situation escalated and retaliatory attacks were carried out by Hindus on Muslims as they sat watching a film at the Sushil theater hall near Godaulia. Muslim men were dragged out of the cinema; three were fatally stabbed and two were burned alive as police watched on. Following these fatal incidents an extended curfew was imposed on the city during which period intensive and intrusive police searches were conducted in Madanpura. It was during one of these searches that a respected sari businessman, social worker of Madanpura and widely known "secular" man, Dr. Anis Ansari, was killed by police when they raided his home (Engineer 1995, pp. 200–203).

The apparent injustice of Dr. Ansari's death, the absence of police account-ability, and the prejudiced reporting of the incident in the local media at the time, continued to underpin contemporary Muslim attitudes towards these institutions. Common to the accounts told by Muslim informants concerning periods of conflict and curfew, was a sense of the severe discrimination meted out by the police. Informants described the brutal and partial treatment they had received at the hands of the police who had conducted indiscriminate searches of their homes, abused their hospitality and looted possessions and properties (Engineer 1995, pp. 203–6; also see Brass 2003).

As a consequence of their tumultuous histories and continuing tensions, these particular religious processions had come to embody a sense of poten-tial danger. They were prepared and conducted with nervous anticipation by both communities, in a way that was sensitive to local and national events. For instance, in 2006, the atmospheres around both these events proved especially tense. Both Durga and Kali Puja occurred during October, which coincided with *Ramzān*; universally observed by Madanpura's Muslims, *Ramzān* represented a particularly conservative and reflective period during the Islamic calendar. Moreover, Kali Puja was coincident with the final night of *Ramzān*, followed the next day by the festivities of Eid al-Fitr.[3] The coin-cidence of Hindu and Muslim festivals in Varanasi has previously given cause for concern. Kumar documents how Durga Puja and Muharram coincided in 1982, heightening the perceived sensitivity of certain localities (1988, p. 213). The convergence of these religious events in 2006 provoked an exceptional demand for Muslim and Bengali communities to cooperate with the administration and to successfully and peacefully manage relations. But, it also lent the GSC club additional leverage in their negotiations with the police about the terms of processing.

In general, expectations that trouble might disrupt the peace functioned to promote the reputation of the *pujā* clubs and local interest in their processions, as both events attracted large crowds from across Varanasi. For example, in 2010 the local media reported that 3,000–4,000 people participated in, or watched the GSC procession (slightly fewer than the 10,000–15,000 people that the Secretary of the GSC had enthusiastically predicted). The violent historical episodes of the processions and ongoing tensions between the Bengali Hindu and Muslim communities ultimately lent the processions a certain degree of notoriety amongst residents in Varanasi. Moreover, for the Puja clubs, the potential for their processions to disrupt everyday peace granted them a degree of local influence and power, which they evidently relished, and actively sought to reproduce, from one year to the next.

Playing with Everyday Peace and (Re)producing Power

Within a context shaped by historical episodes of violent confrontation and curfews, members of the Hindu and Muslim communities alike articulated their desire for the procession to "pass peacefully and without trouble." Yet, ideas about how notions of peace and security should be reproduced and realized in the face of the processions, varied hugely. Whilst local Muslims tended to recognize the perceived potential for tensions to emerge around the processions, they did not generally consider conflict to be a probable outcome, and appeared relatively relaxed in the days leading up to the event. On the other hand, Hindu informants, particularly those closely connected to the *pujā* organizations, expressed a palpable concern that violent conflict was a very real concern, as Sanjay Kumar, the Bengali President of the Nau Sangha described here:

> Personally I always feel that there could be trouble before starting the procession. I always pray to "Ma" [Goddess Kali] that nothing will happen. Speaking truthfully, we all feel like we have a noose (*phandā*) tied around our necks from which we could be suspended at any time.

The President of the Nau Sangha explained his rationale in light of the apparent disgruntlement articulated by the neighboring Muslim community and his concurrent perception concerning the inevitable threat that this might pose to his club's procession:

> There are four reasons: firstly, they (Muslims) don't think that the *mūrti* should go on this route; secondly Muslims think that no fireworks nor *ārti* should be

performed in Madanpura and that we should just go through quietly; thirdly they want the image to be passed through peacefully in their definition of peacefully, but that's not our definition of peaceful. Just simply sitting the *mūrti* on the rickshaw and just going through the area like that is not what the procession is about. Fourthly in the riots of 1991–2 there was a rich man who had contact with Saudi Arabia called Anis, he was the brother of Dr. Mosin and was killed in the riots of 1991–2 by the security forces. So when the image passes in front of the house of Anis, stones are thrown on the procession and the image, always. This procession reminds them of the death of Anis.

These comments reveal the apparent ideological distance that existed between the Puja club and local Muslim inhabitants over the reasonable conduct of the procession. The strong impression, despite "good" relations with certain "respected" members of Madanpura, was that the Nau Sangha President, as well as other club members had made little effort to understand the Muslim community's point of view and resolve tensions. In this respect, both clubs demonstrated egotistical but divergent outlooks about their "right" to process and occupy certain public spaces. I argue that it proved expedient for the clubs to sustain the impression of anticipated conflict in relation to their processions because this not only heightened interest in their respective clubs and their *pujās*, which attracted ever greater donations, but also lent them a certain kind of leverage in their relations with neighboring Muslims, and more importantly, the police. By insisting on holding the *pujā* processions in a manner which most suited them, the *pujā* clubs were intentionally playing with the peace.

Hindu organizations reproduced the notion of anticipated danger by emphasizing the extent to which discussions and decisions concerning the safe passage of the idol were considered in their preparations. The groundwork for the *pūjās* was extensive and involved collecting donations, attention to the style and construction of the idol itself, the lighting and decorations of the *pandāl* in which the image was housed, *pūjā* and hospitality arrangements, and the general staffing of and participation in the event. In addition, meetings with the police administration often took place with different members of the Nau Sangha and the GSC concerning how they should minimize and manage these perceived threats. Prakash and Sanjay Kumar the General Secretary and President respectively, of the Nau Sangh Samiti explained how the senior superintendent of police (SSP), Astoush Pandey, would come for *chāy* and, during the course of this informal meeting, enquire about the schedule for the procession, the *pūjā* and the culture program. The club executives explained how they alerted the police to potentially difficult points of the procession, for example at crossings or where they believed "antisocial elements" wait "to intentionally disrupt the electricity and then throw stones on the image…" The men were keen to emphasize that these meetings were not just with any old police officers but "the big people like the SSP… the DSP (Deputy Superintendent) and SP."

In contrast to the apparently cooperative manner of the Nau Sangha in relation to the police, the GSC deployed their local influence to publicly assert the upper hand with the police. These attitudes were evident during fieldwork in 2006–2008 and were most clearly voiced by Nitin Chowdhury, the GSC General Secretary, during a discussion we had on my return to Varanasi in 2010 as he recalled debates leading up to the procession in the year before:

> Last year a circular went around about the regulations for the worship, preparation and procession of the goddess Durga. But we were not happy, this is our worship and as Hindus we can do it how we want. The administration don't issue circulars for other religions telling them how to practice their worship. They were making all kinds of suggestions, that we should have at least 2 CCTV overlooking the idol, and that there shouldn't be any loudspeakers etc. and, more importantly to us, no "*dhark*" – drums. These are the most important thing in the procession, and something which those in the procession really enjoy.

With no intention of acquiescing to these demands, or making compromises, the GSC called a press conference that successfully put pressure on the police administration to soften the dogmatic tenor of the circular and instead offer "suggestions" for conduct. In renegotiating their position with the police Nitin Chowdhury consciously evoked the potential threat to everyday peace that the procession engendered as a reason why the club should not bend to the calls of the police. For example, Chowdhury contested the request for the procession not to play loud music and to remove drums from the performance by saying that it was the music which kept teenagers occupied during the procession and that without which they were likely to become bored and agitated "and who knows what they'll do, they might start a fight or something … anything could happen." Sure enough the press conference proved successful not least in persuading the police to drop their orders, but also in reproducing the GSC's controversial reputation. Chowdhury undoubtedly reveled in his knowledge that the club exerted local influence as he proudly asserted: "Other clubs are scared of the police, but we're not, we're not servants or animals, why should we be scared?"

The GSC were also involved in something of an ongoing dispute with the police and Madanpura residents involving the construction of *chhajjā* (awning, for protection from rain or sun) above a barber's shop in the *mohallā* which allegedly posed an obstacle to the safe passage of the goddess Durga idol. Raja Dada, a Bengali member of the GSC in his mid-thirties explained the situation:

> RAJA DADA: Once in October 2000 or 2001 next to a *masjid* there was a hair salon where one new *chhajjā* was put up. The owner of the salon was a man called Nizamuddin. We went and complained, saying that the *chhajjā* could bother our procession, in fact so many people complained, but Nizamuddin did not agree to remove the *chhajjā*. So the GSC went in huge numbers. Eventually the administration intervened and decided that the *chhajjā* should be placed higher up, at 24ft. If the statue is 16ft and if it's then carried 3–4ft above the ground, that makes it 19–20ft. In the end the *chhajjā* was pulled down and rebuilt higher up. The administration respected the will of both parties (*Donon ka dil rakh liyā*).
>
> AUTHOR: Who had to pay for it?
>
> RAJA DADA: I think it was the administration who had to pay for it to be rebuilt. Because parties thought the other should pay, so in the end the administration paid for it because they don't want a dispute to arise between Hindus and Muslims.

It is possible that the *chhajjā* was erected to intentionally impede the route, but nonetheless these scenarios point to the apparently close negotiations between the Puja clubs and the police administration. They signal the extent to which the administration more often accommodated the demands of the Hindu community, rather than insisting that they make the compromises. The administration's action in funding the removal of a new balcony illustrated precisely this.

Muslim informants were also conscious of the social muscle exercised by members of the Bengali clubs and their relative powerlessness to resist their intentions. Mohammed Sadiq, a local Muslim resident and small-business owner in the neighborhood, used to be an active member of the *mohallā's* Civil Defense Committee and as a respected inhabitant continued to take an active role in maintaining a safe atmosphere in Madanpura. Despite extensive relations with non-Muslims through social and economic networks, Mohammed Sadiq was less than complimentary about the underhand scheme in which Nau Sangha members had once implicated him:

> These Bengalis are real mother fuckers ... sorry to use this abusive language during *Ramzān* ... To have any religious committee or cooperative you need to have some Muslims and some Hindus on it, only then can it be registered at the government office...The names of some Muslims are represented in these [Puja] clubs but the Muslims are unaware that they are even members. This

once happened to me. It was not in my knowledge that [on paper] I had been made the Vice President of the Nau Sangha. [Before the procession one year] a police inspector came to the committee and enquired as to when the procession would begin. One member of the Nau Sangha committee said: "Our Vice President Mr. Sadiq will tell you, but he is not present at the moment." I understand that one of those present informed the officer where my house was and he called upon me immediately. At the time I was working so I was in very simple clothes and just came out of my house as I was, wearing lunghī and a vest. The policeman asked: "Are you the Vice President? Tell me, what time will the procession begin?" To this I looked at the men in the committee [beside the police inspector] while thinking in my head, you mother fuckers and gave them a piercing look. Then I said to the official: "The procession will begin at 7pm, but right now I have something to do." When the officer left I got really upset with the committee and said: "This is great, so you've become so clever now…" (*yeh bhalā hai…itne chalāk bante hai*).

The widely held perception amongst many in Madanpura was that the Bengali community used the Hindu religious processions, the GSC and Nau Sangha alike, as opportunities to assert and reproduce their power in the locality. The processions offered a chance to play with the peace and test the tolerance levels of Madanpura's Muslims, as Hafiz Habib, a Muslim resident of Madanpura and religious teacher observed:

It [the GSC procession] exists only to disturb the atmosphere unnecessarily. If you see in the past, there was nothing like this 25 or 30 years back. Such things have been introduced to disturb the peace of Hindus and Muslims and create a *dangā* (riot) between us both… at such gatherings you can see all the loafers of both communities of Banaras… all these get together and do everything to provoke the feelings of Muslims. For instance, by lighting fire-crackers in the street which sometimes enter Muslim houses upsetting the Muslims and causing them to become angry and protest. On these days the Muslims keep themselves to themselves and need great control to prevent such things from happening.

As bystanders to the processions, many residents felt they were placed under great pressure to receive with decorum what they perceived as brash and egotistical spectacles. Some Muslim residents I met in the lanes during the Nau Sangha procession had insisted to me that: "You can get a better look on the road, there will be a lot of *tamāsha* and *majmā* ('spectacle and crowd') there!" The particular use of language in this context proved revealing; what would have been seen by some as a sacred Hindu procession, was derogatorily constructed by these bystanders as a flamboyant and entertaining spectacle.

Policing Peace and Making (In)securities

Through their research on UN peacekeepers in Haiti, Liberia and Kosovo, Higate and Henry (2008) have drawn attention to peace as a performance that is constituted in and through space, by particular peacekeeping bodies and their infrastructure. In the days and hours leading up to, and during, the processions, the spaces of Madanpura were entirely reconstituted through the presence of heavy police security, which was drafted in to deter violent conflict and intervene should anything happen. In 2010 for instance, *Dainik Jagran*, a local newspaper, reported that on the eve of the Nau Sangha Puja one company of the Rapid Action Force (RAF), comprising 250 members, six companies of the Provincial Armed Constabulary (PAC) each with 100 men, plus 800 policemen from Varanasi's administration, were deployed on the streets of Madanpura, as well as other police on horseback; tear gas (*ashru* gas) and water cannons were at the ready on the back of fire engines. The visual presence of police bodies and their vehicles in the main high street and lanes, on the rooftops and balconies and escorting the procession along the lanes was overwhelming (see Figure 5.2). The otherwise quiet, peaceful neighborhood had been temporarily transformed into a zone of anticipated conflict.

Figure 5.2   Police presence in the lanes of Madanpura on the evening of the Golden Sporting Club procession. Source: Mike S. Thompson (2006)

It is at the intersection of police bodies, space and place that audiences expressed perceptions about security and insecurity as they assessed the credibility and loyalty of the security force deployed in front of them. As I have suggested, Hindu informants were unswerving about the necessity for high-level police surveillance. Perceiving a real threat posed by Muslims during the course of the procession, and the tacit knowledge that the police administration would protect their interests, local Hindus regarded the police presence as fortifying their security.

Opinions volunteered by Muslims on this matter proved more skeptical about whether security was needed; indeed many regarded the risk of conflict to be low. Rather, concerns more often revolved around the implications of deploying thousands of police in their *mohallā*. Instead of keeping the peace, Madanpura residents were adamant that the high security actually introduced new insecurities. The police presence raised the profile of the processions within the city, which bolstered the self esteem of the Puja clubs and their sense of authority. A middle-aged Muslim silk yarn seller pointed out the irony:

> The security was not needed, but if there was no security then no one would ever find out about it... the security is there to make us see the event, if they don't have such arrangements then how will we know that the GSC image is going for immersion? They take the image through the Muslim area just to show that it's the image of the GSC. Also sometimes because of heated discussions, such instances can convert into riots. This is just to let the world know that the GSC image is going for immersion. There are other ways for the image to go as well, but they choose to pass through Madanpura.

There were also differences of opinion between residents in Madanpura, which reflected their own personal experiences. During the period immediately building up to the GSC procession, I entered into discussion with two men of quite different generations as they drank tea together on Madanpura's high street. We talked about the need for a powerful police presence in the neighborhood:

AUTHOR:	Is security always like this?
ELDERLY MAN:	Every year the security is like this.
AUTHOR:	Is it a good thing?
YOUNGER MAN:	[interrupting] Yes it's necessary.
ELDERLY MAN:	Why did you interrupt like this when a conversation was taking place with someone else? This security creates fear among the communities (*dahshat honā*).
YOUNGER MAN:	There is always this tension, that's why the security is needed to prevent trouble... if a stone was thrown then a *balvā* [trouble, riot] would be inevitable.
ELDERLY MAN:	What's the point of all this [security]? There is no benefit to us.

These views indicate disparities that existed in generational attitudes towards the volatility of local communities during the *Pujā* processions. But the elderly man's final comment interestingly contested the function of the police security and questioned what the Muslim community actually had to gain from its presence. The notion of security does not have an independent reality outside of the social relations that "it" is constituted and sustained. The efficacy of the police and the extent to which they are accepted as legitimate "peacekeepers" is therefore dependent on levels of trust. Implicit in the elderly man's statement is the notion that the police only served the interests of the "majority" community. This sentiment was corroborated by the opinions of Hafiz Habib, an Islamic educationalist and Muslim resident in Madanpura:

> Even the security which is there, it's like the teeth of an elephant, one set to show and the other set of teeth to eat with... If any member of the procession were to hurt a Muslim and so causing the Muslim to retaliate, the police wouldn't take it into consideration that the Muslim had been provoked and so would treat him badly... All politicians in India think that all terrorists are Muslim. But Islam teaches us not to hurt anyone, not to put stones in the path of anyone and to remove those which might hurt others, it doesn't matter whether they are Hindu or Muslim... A few houses down from here, you will get to a Hindu's house, so suppose if there was a riot taking place we would look after the Hindu and give him food. We wouldn't touch him with one finger. But if any one of us was trapped in his area we would certainly not come out alive.

As Hafiz explains, the temporary spaces of insecurity produced through the heavy police presence were underpinned by the perception that the administration's actions were blatantly and predictably partial to its (Bengali) Hindu constituents. The personal encounters of residents with the police and PAC were nested within a wider regional and national awareness of routine police discrimination against Muslims. Concurrently, it was popularly understood by Muslims that the security drafted into the lanes was there to prevent and preempt violence by their community against Bengali Hindus participating in the processions, rather than vice versa. This logic was externally justified and reinforced by local non-Muslim communities; the media and the police who widely portrayed Madanpura as a "sensitive" or "dangerous" area. The upshot of the heavy police presence, drafted in to keep the peace, was therefore contradictory. Whilst it produced perceived security for local Bengali Hindus, the performance of police bodies and deployment of vehicles within their *mohallā* produced new insecurities for local Muslims. As the Puja clubs insisted on continuing to process through their neighborhood, often with few compromises, Madanpura residents were placed on the back foot. Unable to fully trust the state administration to act impartially in maintaining peaceful relations, they were forced to rely on their own, more local initiatives.

Local Peace Initiatives

Mindful of the previously tragic and prejudiced experiences suffered by some residents in Madanpura in connection with the processions, and aware that the police were not guaranteed to protect their interests, informants engaged directly in local initiatives for peace, both formal and informal. These were designed to make sure that conflict between their community and Puja club participants did not arise. During both processions, midway down Madanpura's high street a stage was erected on which members of the Civil Defense Committee occupied a number of tables and chairs. The Civil Defense Committee [*Nagarik Suraksha Samiti*] was a centrally coordinated government organization under the Home Ministry with the official remit to minimize damage to lives, livelihoods and property in the course of disruptions to everyday life.[4] Civil Defense Committees were first established on 24 October 1941; their establishment was later enacted by parliament in May 1968 as the Civil Defense Act 27. The primary intention of the act was to make the States and Union Territories conscious of the need for civil defense action in the face of foreign threats, and to prepare civil defense paper plans for major cities and towns. Resident members of the Civil Defense Committee in Madanpura perceived that the ideological foundations of the association lay in the China–India war of 1962 and were further solidified during the Indo-Pakistan conflict in 1965, as Saful Islam, the Civil Defense Committee Sector Warden explained:

> The Civil Defense (Committee) was introduced at a time of war to encourage the confidence of citizens in their safety and to keep the peace. If there is a war with another country, the Civil Defense's role is to take care of the casualties and dead, and if necessary organize burial or cremation and so on.

Although originally and officially concerned with matters of national security, the distance of a threat of large-scale war in India has meant that the role of the Civil Defense has shifted to more local matters of security, for instance during religious festivals and processions. More often latent in everyday life, the Civil Defense Committee only visibly cohered within public spaces during moments of palpable intercommunity tension. Members of the Civil Defense Committee were nominated by the local sector warden, and their background was then checked by government officials before being officially approved by the District Magistrate. Membership lasted for three years, after which time it was reviewed and could be renewed. Committee members described the review process as the government's strategy to weed out any would-be terrorists. Committee members have to be local residents and educated. Like the police administration, the Civil Defense Committee is organized hierarchically, but it consists entirely of volunteer workers. Then resident in the neighborhood adjacent to Madanpura, Saful lam was the Sector Warden for Madanpura and surrounding *mohallā* and as such, had 250 people under him. The post warden acting as Saful's deputy was Mohammed Zafir, a Shia Muslim who owned a small sari shop in Madanpura and coordinated the efforts of 25 volunteers and took responsibility for reviewing and recruiting new members as needed.

Mohammed Zafir described the role of the Civil Defense Committee during the procession periods as that of liaising with the police and respected members of the Muslim community and in effect acting as intermediaries or mediators between the Muslim community, respective Puja clubs and the administration. He recognized that it was impossible to stop Muslims watching, as they were also keen to witness the events, but he insisted that it was necessary to ensure that "some distance is maintained between the groups by the Civil Defense (Committee)." In the lead-up to the processions, Mohammed Zafir made efforts to negotiate with the police and Puja clubs in order to minimize the disruption caused to everyday life in his *mohallā*. In 2006, Mohammed Zafir appeared satisfied that his concerns had been listened to:

> Last year the procession had came out at 7.45 in the evening, just as the *namāz* of *isha'a* was beginning which meant it came past the Muslim area as *namāz* was finishing. However this year they have shown humanity and will wait until half past eight when *namāz* is completely over.

But, when the procession once again commenced at 7.45pm in 2006 it was evident that the agreement had been contravened by the GSC and again the procession passed through just as *namāz* was finishing. Critically, this time coincided with Muslims thronging out of mosques on to the streets to return home, to visit the houses of friends and family, or to move on to *chāy* stalls. With the main street at its busiest, Hindus and Muslims were compelled to mingle as the procession dominated the vista of the main high street. The intentions of the Civil Defense Committee to design in spaces of separation between Muslims and Hindus had been undermined by the GSC's broken promise to cooperate and respect the Muslim community's request.

Local Madanpura residents further questioned the efficacy of the Civil Defense Committee in maintaining peace. There was a consensus amongst residents that the Civil Defense Committee could do little to effect or maintain conditions of peace, as Imran, a small powerloom business owner and local Muslim observed: "They are only in place to fool people... even if they are just asking a child to move, it wouldn't. Unless we go on to the streets as well, then no one would move when asked." Asked why the Civil Defense Committee commanded little authority in the neighborhood, a prominent local social worker in his thirties, replied that:

> It's the way they explain things, they are rude to the people. But we (local residents) are polite, we explain to people that this is the festival of another community and that we should let them celebrate with enthusiasm.

What explains local residents' lack of conviction in the ability of this more formal association to maintain peace? One answer perhaps lies in the status of the committee members themselves. They were not typically prominent businessmen or widely respected personalities within Madanpura, where status and respect were often equated with financial success. Instead, committee members were more often social workers or individuals who also sought to benefit from occupying an official position within the neighborhood. Many openly enjoyed the privileged mobility and "access to all areas" that their membership granted them during curfew periods. For higher-ranking officials, the Civil Defense Committee also afforded them connections with city officials that could be nurtured in different contexts, for different means. It was evident that relationships with senior police officials brought not only organizational, but also individual, legitimacy. For instance, the Sector Warden, Saful Islam, was expressly proud of his connection with the SSP, Ashutosh Pandey. This sentiment was embodied in Saful's actions on the night of the GSC Puja procession. During the lead-up to the procession Ashutosh Pandey stopped his jeep in the middle of Madanpura's high street and stepped out on to the road, which was by now lined with expectant crowds. Saful Islam confidently strode over to the SSP, shook his hand and held a short conversation about the evening's proceedings. Nearby crowds looked on with curiosity at this relationship between an apparent weaver and a senior police official. After his meeting, Saful returned to the Civil Defense Committee stand visibly excited by the public encounter and the positive impression it must have left with the onlookers.

Yet, the close relationship between the Civil Defense Committee, politicians and the police administration caused many residents to be suspicious of those individuals who joined the committee. A widely held view was that they were *mukbhirs* (police informants) who were given the task of spying on the neighborhood for the state, rather than actually protecting local interests.

For some members, joining the Civil Defense Committee facilitated an improved sense of security for members and their families. Mohammed Zafir lived on the edge of Madanpura, near the boundary with Deonathpura and the immediate locality of the GSC. During the 1977 curfew, the PAC raided his family's home; they broke down the doors, looted ornaments and valuables including savings for his sister's imminent wedding and verbally abused members of his family. Determined to protect himself and his family from future such assaults, especially given his relatively vulnerable location on the *mohallā*'s boundary, Mohammed Zafir strategically joined the Civil Defense Committee. His role on the committee had endowed him with a degree of mobility within and outside the *mohallā* during curfew periods, but more importantly the PAC no longer bothered his family.

Given the well intended, but apparently relatively ineffectual actions of the Civil Defense Committee, in this context it is therefore important to appreciate the role of such formal associations alongside other informal initiatives that were perceived as vital to the process of maintaining everyday peace. Elderly and respected individuals in Madanpura played a central role in working to keep young boys and men out of the *mohallā*'s *galīs* during the procession. On the days of the Durga and Kali Puja processions, shop shutters were closed early and all women and young children were expected to return home. It was clear that local individuals took the responsibility directly upon themselves, rather than entirely relying on the protection of the police and the actions of the Civil Defense Committee. They actively monitored the community throughout the evening to reduce the scope and spaces in which possible conflicts might be sparked by the provoked actions of Muslims. Motivating their energies was a widespread perception that if conflict should result, Muslims would undoubtedly receive the blame. Rafiq bhai, a local Muslim resident and social worker, expressed the irony of their endeavors:

> If we're talking about the GSC procession then yes, I always go there to see the processions, we are all there, but we have a purpose on the road, not to watch the procession, but to make sure that nothing will happen. If the peace is destroyed by someone else then it's always the Muslims who will be blamed for it. The action is very one-sided, I'm not blaming one community, it's the fact.

What caused respected members of the Muslim community particular concern was the presence of youths or "antisocial elements" from other neighborhoods who came into the *mohallā* for the procession. Curt words and a determined dismissal were delivered by elderly inhabitants to some unfamiliar teenage boys who were hanging around the *mohallā* lanes, curious to see what might happen on the night of Kali Puja. Inside the *mohallā*, at the junction where the route of the GSC's procession had been contested there was a heavy presence of local Muslim men and police. A rope and bamboo barrier cordoned off the route, behind which, policemen in riot gear formed layers three deep across the narrow lane and behind them middle aged as well as more elderly Muslim male residents made up a further four or five layers. The back row was keeping vigilance for troublemakers and frequently encouraged young boys to move away from the lane.

An informant noted that on occasions like this, the proximity and manner of the religious procession, in bringing the Hindu community directly through the Muslim neighborhood embodied a distinct provocative potential. As a local Islamic teacher and respected character in the neighborhood pointed out, this demanded that close attention be paid to the actions of young inhabitants in the *mohallā*:

> HAFIZ HABIB: At the time of the of the procession we tell our young people to move away from the procession, or if they are present then we advise them to be controlled and not to react to any abuse. It's like this, suppose there are two live electric wires, if they both touch then the bulb will fuse...
>
> AUTHOR: You talk about these live wires, would you say they are live in everyday life?
>
> HAFIZ HABIB: They are not live in everyday life, only during that (procession) time. But those people (Hindus) throughout the year are heated towards us. Our religion teaches *mohabbat* (love, affection), if there's no *mohabbat* there's no *īmān* (truth).
>
> AUTHOR: Why are they always so heated towards you?
>
> HAFIZ HABIB: ... [Pause] ... we don't know why they think like that, they can tell you, not us... They say that India is not the country for Muslims. Higher politicians mislead them and say that this is not the country for us Muslims and that they should chase us away from here and kill us.

These resonated strongly with the opinions of others in Madanpura and thereby drew attention to the simultaneity of tension and distrust that shaped the actions of local Muslim residents in the maintenance of everyday peace.

Conclusions

In 2010 on the morning after the Golden Sporting Club procession, *Aaj*, the local Hindi newspaper reported that the idol was immersed "peacefully and excellently, resulting in peace and happiness for the police administration along with those participating." Such talk of peace acts to reaffirm peaceful relations in the city and in place, but it simultaneously perpetuates the notion that everyday peace is a fragile process that cannot be taken for granted. This chapter has extended perspectives on peace by pointing to the uneven geography of peace as process, the intricate relationship between violence and peace and the contrasting influence of local peace initiatives and agency.

The chapter firstly advances conceptualizations of peace as process by highlighting the uneven geographies of everyday peace which make it vulnerable to subversion by certain groups with the potential to mobilize political capital. Memories of previous intercommunity violence in the neighborhood during religious Hindu processions were folded into the everyday, and where tensions remained suspended and unresolved they underpinned a precarious kind of peace that left open the potential for future tension and conflict. Local Bengali Hindus reproduced their power and influence in the locality with respect to the local Muslim population and police by challenging the robustness of everyday peace and exploiting the fragility of peace. In taking their *pujā* processions through Madanpura, the Bengali clubs strategically and openly played with peace by evoking histories of violence and difference in order to bolster their local influence and authority, especially vis-à-vis local Muslim residents. During these annual events, the nature of encounters between local Bengali Hindu communities and Muslim residents was confrontational and oftentimes intimidatory. Where Bengali Hindus asserted their right to process visibly and loudly through Madanpura's narrow lanes, this act of playing with peace constituted forms of political capital for the Puja clubs and their members. The uncertain outcome of these encounters sustained low-level tensions between neighboring communities which undoubtedly constrained how Muslims chose to articulate their citizenship, as I outlined in the previous chapter.

Second, further developing a view on the uneven geographies of peace, the chapter documented a situated account of how, in practice, keeping the peace meant different things for different people. The sense of hype and threat to everyday peace that the processions embodied served to justify the considerable presence of police personnel and equipment in the neighborhood. Yet, these state-imposed "peacekeepers" constituted conflicting notions of security and insecurity for different groups. For the Bengali club members, the police, PAC and RAF contributed to producing an atmosphere of security, but for local Muslims they actually created new spaces of insecurity and served to reinforce wider perceptions concerning their marginalization as a minority within Indian society.

Thirdly, in contrast to Chapter Three, which emphasized the agency of community leaders in maintaining peace, this view on everyday peace documented the relative influence of local residents and civil society in protecting peaceful relations. The chapter develops an understanding of formal and informal mechanisms for peace within civil society, in conversation with the work of Ashutosh Varshney (2002). The public appearance of the Civil Defense Committee around the time of these particular *pujā* processions, contrasts with the manifestation of "peace committees" that Varshney (2002) documents, in Lucknow and Hyderabad for instance. He determines that "peace committees" in Lucknow were merely an extension of pre-existing local networks of engagement between Hindus and Muslims. These emerged at times of tension out of routine economic interactions and consequently proved effective at maintaining peace. Conversely in Hyderabad, the characteristic low-intensity of Hindu–Muslim interactions meant that "peace committees" did not generally materialize organically and neither were associations effectively implemented "from above."

Whilst the economic interests of Madanpura's Muslim residents were extensively interlocked with those of Varanasi's Hindu merchants (largely based in Kunj gali), the Bengali Hindu neighbors in question were not typically involved in the silk sari industry, instead preferring professional occupations. The Hindu–Muslim ties that were nonetheless present were developed in social rather than economic spaces, between neighbors or acquaintances over the course of years as they encountered each other in everyday social spaces. Given the degree of cooperation within the *mohallā* around security issues, intra-community ties (those within the neighborhood) proved especially important in this scenario. Since economic interactions did not constitute the backbone of intercommunity relations in this instance, it is less surprising that a local "peace committee" did not organically emerge as a natural extension of everyday economic engagements (see Varshney 2002) and associational engagements did not therefore constitute the primary mechanism for maintaining peace. In order to explain why and how everyday peace was maintained in this context, it is therefore important that the work of formal mechanisms, such as the Civil Defense Committee, is viewed through a wider lens that also includes other more informal and ad-hoc initiatives for peace.

Where the Civil Defense Committee was perceived to lack authenticity and efficacy in the neighborhood and police administration was discriminatory, local Muslim residents felt a level of responsibility to provide alternative and more authentic spaces of surveillance. Accordingly, different Muslim residents implemented a series of local and ad hoc initiatives for peace. In particular, respected local Muslim males were dynamic in the surveillance and management of people within the lanes of the *mohallā*, especially younger Muslim youths, some of whom were believed to come from outside the neighborhood looking for "excitement."

This chapter demonstrated how experiences of unequal citizenship contributed to, and reinforced, contrasting experiences of peace and security and underpinned the mobilization of informal local initiatives to maintain peace. The next chapter shifts the focus to the city's silk sari market to examine how the type and quality of everyday interaction and encounter between Muslims and Hindus reproduces forms of everyday peace.

The uneven geographies of peace and power become particularly apparent in this economic space, which underpins one of the book's main arguments, that we need to conceive of peace itself as political.

Notes

1 Durga, Saraswati and Kali Pujas coincide respectively with Dashera, Basant Panchami and Diwali in the UP festival calendar.
2 Dussehra is celebrated on the tenth day of the Hindu autumn lunar month of Ashvin, or Ashwayuja which falls in September or October of the Western calendar. It comes after nine days of celebrating Maha Navratri and represents the victory of good over evil.
3 Muslim holiday that marks the end of *Ramzān*, the Muslim holy month of fasting.
4 In Madanpura, the Civil Defense committee was sometimes locally referred to as the "peace committee" however there is a clear distinction between the two in real terms. I shall therefore only use the term Civil Defense committee. "Peace committees" on the other hand represent the more organic and locally coordinated groups that have emerged during riots in the city. Madanpura residents described how such "peace committees" had disbanded in recent years as a consequence of the infrequency of riots.

Chapter Six
Economic Peace and the Silk Sari Market

The silk sari market represents an important site for understanding everyday peace in Varanasi, for it is where practices of *bhaī-bhaī* or *bhaichārā* (brotherhood) and interdependence between the city's Hindus and Muslims were most tangible and most talked about. As Chapter Three documented, the possibility for restoring peace in the aftermath of the terrorist attacks was in part credited to the presence of economic networks between Hindus and Muslims which engendered both real and imagined pathways to intercommunity peace. However, as Chapter Four and Chapter Five have suggested, Varanasi's Muslims experience patterns of perceived discrimination by the state and sections of Indian society. This experience of unequal citizenship underpins peaceful realities and attitudes towards peace across different sites and scale. In this chapter I hold in mind these wider contexts of disadvantage and examine the nature of interactions that take place in the economy between Muslims and their Hindu counterparts, and how these are constitutive of everyday peace.

The relationships between religious groups in the economy and how these impact on broader experiences of everyday peace, or not, prove a pertinent site for investigation. The interacting politics of the state's secular aspirations and the everyday existence of religious imaginaries and realities has led to a situation

Everyday Peace?: Politics, Citizenship and Muslim Lives in India, First Edition. Philippa Williams.
© 2015 John Wiley & Sons, Ltd. Published 2015 by John Wiley & Sons, Ltd.

in which the Indian economy has been open to the influence of religious identities and communities. In the informal economy especially, this has given rise to religiously based segmentation and coexistence as well as competition, ranging from minor disputes to more fierce antagonism. Competition along religious lines inevitably reinforces the conditions which make religious organization necessary and hierarchies inevitable (Harriss-White 2005).

Scholars have examined how social relations are mediated within different work settings and how repeated and intimate contact, shared livelihoods and localized practices develop *in situ*. Studies inside American firms have sought to understand how cultural diversity enhances or detracts from the functioning of groups within office environments (Thomas and Ely 2001) as well as how workplace bonds may strengthen democracy in diverse societies (Estlund 2003). South Asian research has touched upon how different religious and ethnic groups encounter and engage with one another in work and trade settings (e.g. Mines 1972; Kumar 1988; Lau 2009). Others have shown how patterns of trust, affection and friendship differentially intersect and inform working relations (Harriss 1999; Knorringa 1999), and how these dynamics are critically shaped by sociopolitical events at different scales (Froystad 2005). Studies that have touched upon Hindu–Muslim friendships in north India have shown that relationships centered on mutual exchange (*len-den*) and reciprocity were those most likely to endure over time (Froystad 2005, Ciotti 2008). The focus on intercommunity spaces of work and the reproduction of everyday peace demands an attention to the ways in which Hindu–Muslim encounters and transactions are situated within networks of value critically mediated by money (see Chari and Gidwani 2004). The circulation of economic capital is therefore integral to the possibility, as well as the necessity, of everyday peace being sustained within the silk industry. A shared desire for money may act as an incentive to maintain amicable, peaceful relations. But, how existing networks of value are held together and new ones produced is also contingent on disciplinary and occasionally violent operations of the market, involving for instance, coercion and consent, the exercise of surveillance, regulation, repression, exploitation and primitive accumulation. Examining everyday peace through the silk sari market therefore requires a close concern for the (potentially) violent aspects of capital as well as the acts of fabrication that sustain and shape the industry and its constituent intercommunity relations.

In this economic context, the focus on Muslim weavers therefore draws attention to the uneven geographies of power that inform and are perpetuated through the reproduction of peaceful market relations. And, it reveals how these inequalities are bound into relations of everyday peace through practices of surveillance, exploitation, coercion, consent and complicity.

This chapter therefore develops a perspective on the everyday encounter and the politics of peace, and draws attention to the role of the political within peaceful spaces, something which has been underexplored in current literature (e.g. Varshney 2002). The chapter is structured in four sections. First I document the imagined spaces of intercommunity peace that find expression through forms of "peace talk" that were originally evoked in the Introduction to the book. Second, I examine "peace at work" to contrast and critique these imagined forms of peace in light of Hindu–Muslim working realities. This section shows how everyday encounters are constituted through uneven day-to-day exchange, profit incentives and differentially shaped by notions of friendship, acceptance and indifference. In the third section I illustrate how peace is, in and of itself, political; this becomes especially apparent in the spaces of the sari market and, as I show in the fourth section, where notions of difference shift through space and time and have to be actively negotiated and reproduced. The final section considers how economic transformation amongst Muslims may be realized in the context of maintaining Hindu–Muslim peace, and the different kinds of agency, such as learning, pragmatism and resilience, which are articulated by local Muslims looking for more prosperous and peaceful futures.

Imagining Everyday Peace and "Peace Talk"

The silk sari market in Varanasi is widely talked about as a site of intercommunity harmony in north India: local, national and international discourses celebrate the city as a paragon of Hindu–Muslim collaboration. So universal and unremitting were informants' narratives that spoke of mutual dependence and cooperation it sometimes appeared as if a preordained script existed that was not to be deviated from. Given the niche occupations of Hindus and Muslims within the industry their vertical, and to a degree horizontal, interdependence was central to the successful production and reproduction of the industry.

Interactions between religious communities were often analogized to the process of weaving itself; "they are like '*tānā bānā*' (or 'warp and weft')" was a particularly popular characterization. Silk fabric is comprised of interwoven horizontal and vertically placed threads; the *tānā* or warp sits on the loom while the *bānā* or weft is woven through the *tānā*. Just like silk, the integrity of the market was purported to be contingent on the quality of Hindu–Muslim relations. Meanwhile, others extended the metaphor to refer to the way a sari and its components are worn describing them as "*choli dāman ka sāth*" (like with a women's blouse and sari hem). Here, they compared intercommunity bonds to the intimate proximity of a women's blouse and the furthest hem of the sari. In both cases the analogies reflect the importance of their dependency on each other while recognizing their differences.

More than recognizing the necessary intimacy and cooperation between Hindus and Muslims in the workspace, informants also spoke of the presence of "Hindu–Muslim brotherhood" articulated in terms of *bhaī-bhaī* or *bhaichārā* (brotherhood or brotherly share) (see also Hussain 2008 and Jeffrey 2010). This term was evoked to describe the mutual affection, trust and loyalty that existed between Hindu and Muslim parties and pointed towards their equality. Perhaps at once recognizing and pre-empting more popular imaginaries about Hindu–Muslim antagonism, informants often further qualified the possibility of intercommunity brotherhood in the context of regional cross-cultural collaborations, for example in music and dance. As in the aftermath of the terrorist attacks detailed in Chapter Three, the notion of brotherhood was connected to the region's cultural economy, embodied in the notion of *Gangā Jamuni tahzēb/sanskritī*.

Narratives of peace may both construct and maintain the status quo. In the Indian city of Ajmer, Mayaram describes how the city's tradition of "communal harmony" is evoked as a narrative trope (Mayaram 2005), which at once reflects a reality but also impacts how different groups defuse rather than aggravate tense situations. Further illustrating the role that discourse plays in constructing everyday peace, Carolyn Heitmeyer (2009) focuses attention on everyday living together in the Indian state of Gujarat three years after widespread anti-Muslim violence played out across the region. In the small town of Sultanpur, she observes that, unlike in nearby cities such as Ahmedabad and Vadodara, Hindus and Muslims continued to coexist and to forge close intercommunity relationships even in the face of the violence. In the course of these interactions, notions of Hindu and Muslim coexistence were constructed and reinforced through various "normalizing" discourses that functioned to externalize acts and agents of violence from Sultanpur life. Despite the town being paralyzed for more than two months by riots, residents nonetheless persistently maintained that "*Sultanpurma shanti che*" ("there is peace in Sultanpur"). Ordinary citizens routinely reiterated the discourse of "communal harmony," and local residents took great pride in their town as a haven from the noise, pollution, and communalism of the larger urban centers nearby. In so doing, these discourses rhetorically restored notions of peace within everyday life. But what is less clear from these studies is how different individuals or communities are positioned vis-à-vis these narrative constructions.

In this context, "brotherhood" traditionally suggests the existence of equal respect, through virtue of being of the same father. Through a sense of this shared fate, feelings of affection, trust and loyalty may arise. But, as a natural and inevitable relation that contrasts with friendship nurtured more often through choice, brothers are rarely equals, whether in terms of age or acquired characteristics, such as strength or wealth (for a discussion on friendship, see Dyson 2010). Competition, contest and conflict are therefore typical traits of brother-like interactions. The idiom of "Hindu–Muslim brotherhood" at once signals a degree of mutual trust and respect given their shared economic fate but it also recognizes the divergent religious identities and conflicting histories. Indeed, their shared violent histories to an extent facilitated or even necessitated the active reproduction of brotherhood. "Hindu–Muslim brotherhood" therefore represents an imagined geography of peace whilst simultaneously constituting the realm of the political and the real possibility of transformation into enmity (Kapila 2010).

It should be noted that the notion of "Hindu–Muslim brotherhood" suggests an exclusively male domain. Indeed, the interaction and transactional spaces were almost entirely dominated by men, with the exception of some female Hindu consumers. But, inside Muslim homes, female input in the industry was ubiquitous and extensive in scope. From keeping business accounts to spinning yarn, preparing bobbins and cutting excess *zarī* as well as household management and childcare, females of all ages were the silent backbone of the industry. Muslim women may have been conspicuously absent from the public spaces of interaction that constitute the marketplace and which represent the focus of this chapter, but they also invested in this imaginary. Moreover, female and child labor in the home aided the material reproduction of male Hindu–Muslim interaction in the market. Just as war discourses are gendered (see Dowler 2002), women and children appeared absent from constructions of peace, even while they actively contributed towards its reproduction (see Kumar 2007).

Similarly, in Varanasi discourses of peace played a powerful role in reproducing everyday realities. Even if not experienced as an intimate everyday practice by all, the narrative construction of Hindu–Muslim brotherhood functioned as a normalizing discourse which denied the existence of conflict in the silk sari market and reproduced notions of peace in the city more widely, as Chapter Three documented. Following Heathershaw's (2008) argument with respect to popular Tajik discourses of peace, I propose that this vernacular imaginary of peace proved productive not because it denied the existence of inequality or tensions, but because it acted in some ways as a control over what is and is not acceptable in the public space, and therefore functioned as a self fulfilling prophecy.

Peace at Work

In this section I explore further how notions of "brotherhood" and mutual dependence were actually experienced and articulated by different people, by examining the working realities of intercommunity relations in the silk market and Muslim-majority neighborhood. In thinking about the nature of encounter,

I suggest, firstly, that vernacular narratives of brotherhood sustained the idea of cooperation marked by the notion of equal respect and shared affection. This was possible, in part by distracting from the inherently unequal nature of relations and practices of exploitation through which intercommunity interdependence was produced and reproduced. Secondly, I draw attention to the centrality of capital in sustaining Hindu–Muslim intimacy and third, I document contrasting forms of encounter and the differential ways in which friendship, acceptance, trust and indifference shaped transactions.

Uneven Exchange

More often, Muslim weavers were the more financially vulnerable, and experienced widespread practices of exploitation and discrimination within the industry. The challenging predicament of Muslim *Ansāris* in Varanasi has been widely documented by academic as well as popular and policy writers (Kumar 1988; Showeb 1994; *The Economist* 2009). Here I consider just some of the issues faced by weavers that also had implications for the reproduction of everyday Hindu–Muslim peace in Varanasi. Hindu traders were renowned for keeping their weavers, who were usually Muslims, suspended in positions of financial limbo and economic dependence through the practice of delaying payments on the one hand and advancing loans on the other. In addition to the perils of economic entrapment, weavers had not experienced a rise in their wages and had no choice but to absorb increases in the price of raw materials. A typical view I heard from weavers was that "they [traders] cut off our wages and continue to build their houses." Traders also frequently found excuse to deduct a fraction from the weavers' wage upon finding fault with their product, however minor. Through a range of exploitative practices traders maintained their dominance such that the poorest of weavers were compelled to sustain relationships often through coercion rather than choice.

Broader difficulties in accessing credit also inhibited Muslim weavers' opportunities for upward mobility within the industry and their concurrent freedom from Hindu traders and brokers. As Altaf, a teenage Muslim weaver, spun yarn on to the *ante* (large loom for ordering silk thread) in his uncle's workshop, he told me about his father's unsuccessful attempt to obtain a loan from the bank:

> Taking a loan is not that easy; it took us six to seven months. We couldn't just go straight to the bank because we don't know how to get things done, we're not educated. So my father went to a middleman who was from our own *birādarī* (community), he wanted a bribe which meant we would only get half the loan we applied for: Rs25,000 instead of Rs50,000. But the middleman cheated us somehow so we never actually got the loan, and yet we're still paying it back.

Few weavers had been educated past grades three or four (i.e. up to 8–9 years of age). A widespread lack of education encouraged a culture of ignorance and suspicion around available banking facilities. Consequently, weavers' autonomy in accessing advance capital was severely hindered and their subsequent dependence on middlemen left them vulnerable to further exploitation in the form of heavy bribes (see Chari 2004). The prevalence of stories about neighbors or relatives who had found themselves swindled by bankers and middlemen only heightened a self-perception of vulnerability vis-à-vis majority Hindus. While low-caste Hindus undoubtedly suffered similar issues of access, Muslims in particular perceived discrimination within banking institutions through the lens of their religious identity. These economic conditions, therefore, structured the environment within which processes of peace were imagined and realized.

Where a small stratum of the urban Muslim community had managed to consolidate their role in business it is worth reflecting on the continued dominance exerted by Hindu traders who resided in Chowk. In the mid-1970s attempts were made to establish a weavers' market in Badi Bazar in Varanasi with the express purpose of circumventing the middlemen at Chowk and bringing weavers greater profits. The market survived little more than a year. It was alleged that traders at Chowk reacted to the new competition by actively raising their buying price per sari from Rs200 to Rs210 in order to maintain business with weavers and consequently hinder the economic growth of Badi Bazar. In 2007 the purpose-built space partially operated as a *dhāgā* (thread) market instead. Hindu traders frequently recounted narratives around the "failure of the weavers' market" as evidence of the industry's inherent competition and the incapability of Muslim counterparts to keep pace. Behind the scenes however, it was evident that Hindu traders flexed not only considerable economic, but also political, muscle within the city through their participation in trade associations and the grassroots party politics of the Hindu right.

Capital Motive for Intimacy?

> There is always a need for profit... there is a mad rush for profit, to squeeze the profit. At the end of the day, the status of the buyer doesn't matter, whether they're Hindu or Muslim, it always comes down to the best profit.
>
> Professor Toha, Retired sociology professor and resident of Madanpura, November 2006

The primacy afforded to profit, or simply payment for labor in the case of weavers, proved a very strong force in binding intercommunity transactions, as this local Madanpura resident and retired professor observed. In the drive for profit it appeared that religious identities, and any potential differences that these might have presented, were suspended. A range of considerations and conditions mediated how different individuals sought to secure their financial capital, whether in the form of daily wages for weavers or gross annual turnover for businessmen. But in the course of routine business what was more important to informants was the monetary value, promptness of the transaction, and sometimes the trustworthiness of the individual concerned. The case of Nadeem, a Muslim middleman is illustrative. Nadeem operated as an agent collecting polishing and dry cleaning work from Chowk to distribute amongst workers in his majority Muslim *mohallā* (neighbor-hood). Over 20 years mediating between the spaces of the market and *mohallā* he had developed a specialist business network in Chowk. While Nadeem preferred to work for cordial traders, the need for capital took precedent as he explained with particular reference to the *gaddidār* at T.P. Saris, in Chowk.

> I wouldn't work for him by choice, but it's only because of my stomach [hunger] that I have to... Whenever I go to him for work I have to beg and behave so humbly. But, although his manner is very bad he pays very well, on a monthly basis, while others pay after three months or longer.

Conversely, Nadeem lamented how Soni Jaiswal, a very good-natured and polite Hindu *gaddidār*, did not have the volume of work to support him. He regretted the paucity of business interactions with such a likeable man. Nadeem preferred to return to those dealers with whom he had developed relations and knew something of their reputation, through his or others' experiences. Business relations with traders were therefore privileged on the basis of capital returns and trustworthiness; while conviviality was desired it was less influential in sustaining relations. In a setting where getting business done mattered most, Nadeem was indifferent to the religious backgrounds of his respective clients.

Acceptance, Friendship and Indifference

On the one hand I have shown how religious identities mattered little on a day-to-day basis in a market context where the reproduction of economic capital was the primary concern. And on the other hand I have described the stratified structure of Hindu–Muslim relationships that critically informed the working realities of everyday peace and underpinned the conditions for the existence and persistence of intercommunity economic ties. Within these settings, informants articulated their position and relations vis-à-vis the other community in a range of ways, which, during my period of fieldwork, were orientated towards the maintenance of peaceful exchanges.

On the nature of intercommunity interdependence Shakil Ahmed, an elderly Muslim weaver candidly reflected: "It's not helplessness; it's the *mohabbat* (love, affection) which keeps us together. But you could call it help-lessness as well." This "helplessness," rooted in their economic inequality, coerced and tied weavers into frequently exploitative relationships, which in turn contributed to the heightening and persistence of their (economic) vulnerability. In this respect, the unbroken reality of intercommunity relations was vital to their very survival. In the context of this economic exploitation, Freitag suggests that the persistence of peace may be regarded as a "desperate kind of harmony" (1992, p. 168). There have been occasions when weavers refused to participate in a communal quarrel of larger provincial dimensions because "our *rozi-roti*" (livelihood) depends on the Hindus" (*Bharat Jiwan*, 27 May 1907, p. 8 cited in Freitag 1992, p. 168).

Some Muslim weavers attributed their marginalized situation to "*kismat*" (fate, fortune) and employed this notion as a means of explaining and rational-izing their apparently defenseless position. To an extent this narrative served to maintain the dignity of its weaver-author, since the possibility of transforming his condition was not something he could reasonably influence. In some respect conceiving of oneself as a vital connection in a wider critical network offered a way of coping with the everyday. But, at the same time accepting one's position within that network inevitably entailed a degree of consent to exist within violent economic relationships (see Connolly 1979). That weavers demonstrated everyday forms of acceptance offers an interesting counter to Scott's (1985) notion of everyday acts of resistance thereby broadening our understanding of subaltern agency and the mode and possible extent of trans-formations, where structures of power go apparently unchallenged.

Whilst many poorer weavers like Shakil Ahmed did not report having friendships with Hindus in the industry, for others, such as Nadeem, everyday intercommunity brotherhood was a lived reality. Not only did he conduct repeat and routine interactions with Hindu businessmen, but he also frequented their homes, for instance while visiting a mosque on the outskirts of Varanasi with his wife from time to time, or during religious festivals. Such friendships were not uncommon in Varanasi, where a sense of shared Banarasi culture and pastimes underpinned intercommunity friendships within and beyond the industry (Kumar 1988). Indeed for Nadeem's line of work it was important that he invested considerable labor in building trust and nurturing respect with (mainly Hindu) traders, in an effort to win over and sustain their business. But where friendship and trust were not always prevalent features of intercommunity exchange it is important to note that the civility of indifference (Bailey 1996) more often played a vital role in reproducing cosmopolitan encounters (Nielsen 2008) in the marketplace.

These empirical insights into the nature of everyday exchange and encounter between Muslims and Hindus collectively reflect a plurality of experiences and motivations for sustaining peaceful relations. They therefore disrupt otherwise universal imaginations of brotherhood and point to embedded and uneven geographies of power and difference which also constitute peaceful realities, and which I turn to next.

Peace and the Political, Reproducing Difference

Despite the apparent inconsequence of religious affiliation in business transactions within and between the Muslim neighborhood and sari market, the dominant narrative of Hindu–Muslim brotherhood was constructed on the basis of apparent Hindu–Muslim distinction. In turn, the active deployment of this narrative acted to reproduce notions of difference, even while connections and friendships were asserted. From the loom to the market, such distinctions were differentially embodied, evoked and experienced through material and imagined encounters. Experiences of difference within the sari market shifted through space *and* time, informed by events in the neighborhood, city, region and nation.

Muslims and Hindus most frequently encountered one another in the main sari market at Kunj gali as they passed in the lanes, drank *chāy* from the same tea stalls, bought and sold saris from one another and touted for new business. In these spaces especially, different parties were visibly distinctive both in terms of their appearance and the spaces they occupied. Hindu traders reclined on the bolsters of their shops, which were raised above the narrow lanes. They often sported a vermillion *tilak* on their foreheads which signaled their recent participation in Hindu worship, while paintings of Hindu gods adorned their shop walls (see Figure 6.1). Meanwhile, Muslim weavers moved along the lanes between the shops, clutching saris wrapped in cloth; their style of *lunghī* (cotton cloth tied around the waist like a skirt) or white *kurtā-pyjāmā* (knee length shift over straight trousers) in combination with a *dārhī* (beard) and *topī* (small cap) marked them out at once as Muslim and as weaver. Filling an in between space, young commission agents, mostly but not exclusively Hindus, wore trousers and shirts as they moved between traders, buyers and weavers (see Figure 6.2).

Figure 6.1   Hindu traders wait for business in Kunj gali, Chowk. Source: Vinay Sharma (2007)

Figure 6.2   Intercommunity encounter in the lanes of Kunj gali, Chowk. Source: Vinay Sharma (2007)

The notably different form of dress worn by Muslim weavers represented a basis for routine and negative stereotyping by Hindu traders. Muslim middlemen who conducted business with Hindus and Muslims were often conscious of their embodied and marginalized identities and commonly opted to wear mainstream "Western" attire, comprising shirt and trousers instead. It was inferred by informants that by resembling the majority community their "Muslimness" was more ambiguous and, consequently, their mobility within and between industry localities made easier. Moreover, some informants suggested that should tension or conflict arise between religious communities whilst they were in a Hindu-majority area they were likely to feel safer and at less risk of being targeted.

The imbalance of power in "Hindu trader"–"Muslim weaver" relations found expression in narratives of frustration that were grounded in a feeling of Muslim inferiority vis-à-vis Hindu dealers. A retired weaver and the father of a middle-class *gaddidār* in Madanpura described the aggravation that the Muslim weaving community had experienced and their continued resilience in the business. This was in spite of their ongoing setbacks in the industry and heightened experiences of religious and occupational discrimination in recent decades.

What proved interesting through observing and interviewing Hindu traders based in Kunj gali was that notions of difference articulated and perpetuated through stereotyped constructions of the Muslim "Other" existed alongside everyday practices of intercommunity interaction and exchange. The opinions of Dinesh ji, a young Hindu *gaddidār* from a family of silk traders, are broadly representative. He purchased a variety of goods from Hindu and Muslim weavers across the city's neighborhoods. Discussing his experiences working with weavers from different communities Dinesh ji made clear distinctions between Hindus and Muslims:

DINESH JI: They [Muslim weavers] are useless. What a Hindu can understand a Muslim doesn't. Whenever they come with a sari, even if it's just one, then they want payment right away. Their behavior is always different [to Hindu]. If you give an advance you will never get it back from a Muslim... But if I go to get the money the person will say they have no stove, no food, but at the same time they're living in a four-storey building.

AUTHOR: But, why do you think there is a difference (between Hindus and Muslims)?

DINESH JI: Hindus understand (a *gaddidār's* situation), while Muslims don't. They are uneducated, they weave day and night and the next morning they know they have earned Rs50, so they want the money right away. This misunderstanding could be due to poverty, education, big families... Muslims usually have six or more children; they are very talented at making children... They have to feed them so that's why they need money every day.

These discourses of suspicion and mistrust are situated within a stereotypical and negative portrayal of the Muslim community. Despite his broadly denigrating attitudes towards Muslim weavers, Dinesh ji nonetheless emphasized the strong relations between him and his weavers, which he verified by his frequent participation in their weddings. Describing the last wedding he attended, Dinesh ji was careful to point out that he visited just to deliver a gift and then returned home without consuming the food provided, because he was vegetarian. While supporting and building his business relations across religious identities, in the act of not taking food or prolonging his stay, Dinesh ji simultaneously reconstructed the social boundary between his Hindu identity and his weaver's Muslim identity.

The reproduction of everyday economic peace takes place within a setting where community differences are recognized, negotiated, accepted and reproduced in a variety of sometimes contradictory ways. The important point is that in spite of perceived differences, everyday actions of both communities are orientated towards maintaining the status quo, everyday peace, and concurrently the conditions necessary for the circulation of economic capital. Yet, as the next section shows, the quality of everyday peace shifts through space and time, contingent on wider socioeconomic and political events.

Spaces of Contest and Cooperation

In Varanasi it was evident that over recent decades everyday peaceful working realities had undergone a perceived shift following the highly publicized Hindu–Muslim riots in India. The lingering sense of fear, insecurity and injustice that these incidents provoked amongst both communities, but especially India's Muslims, informed the background against which Muslim weavers and traders performed daily transactions with other communities. This reality contrasts with a tendency to imagine situations of peace as unchanging and as the aftermath or absence of violence. Instead, as I show here, the character of peaceful relations transformed through space and time.

From 1989 until 1992 the national and international focus was on a disputed mosque at Ayodhya. Hindu nationalists claimed that a particular site in Ayodhya, UP, represented the birthplace of the revered Hindu god, Lord Ram and accordingly the Babri mosque erected on that same site in 1527 should be demolished. On 6 December 1992 such rhetorical posturing transformed into violent action as young Hindu males demolished the mosque. The act of demolition provoked tension between Hindus and Muslims across north India and led to Hindu–Muslim riots in a number of urban areas (Engineer 1995). During this period Varanasi witnessed its most serious intercommunity violence (Malik 1996). Such events inevitably had a direct impact upon relations in the silk industry and the nature of everyday peace, as a retired Muslim weaver reflected:

> The riots had a very bad affect on this profession... Although relations are generally good [with Hindus], when anything happens [i.e. a riot] the relations turn a bit bitter for a while and a distance is created before coming back on track. But businessmen don't keep such things in mind and amongst the Hindus not all are bad. Ten in one hundred are bad, most people do not want riots to happen... I'm not saying that all Muslims are good, they are bad too!

Narratives concerning incidents of riots often described a typical period of temporary hostility manifest in a physical distancing between the two religious communities immediately after an incident of violent conflict. The main sari market at Kunj gali and shops in Muslim areas would close for a few days and workers would stay away waiting to see whether tensions would develop or dissipate. It was commonly accepted that periods of riots and curfews had been hugely detrimental to the sari business and were therefore not in the interests of any party.

During 1990s curfews were not uncommon, particularly within so-called "sensitive areas" that were understood by the police to be especially prone to "trouble." As elsewhere in India, the city's majority Hindu police administration perceived that the greatest threats to peace emanated from Muslim *Ansāri* neighborhoods. Even in the early twentieth century, Gooptu shows how Muslim *Ansāris* were regarded as a menace to the health and prosperity of peace in Varanasi (2001, p. 261).

This approach was informed by and reinforced narratives that openly circulated about Muslims as "dangerous," in possession of weapons and "anti-national" (Hansen 1996). Muslim *Ansāris* were familiar with this prejudicial treatment and were resigned to accept or at least tolerate the situation, often through humor. Muslim informants regularly made ironic references to their neighborhood as "mini Pakistan" and joked that during curfew periods they answered the telephone with "*curfew mubarak ho*" or "happy curfew" every time another new curfew was imposed on their area. On a more serious note, however, curfews were particularly detrimental to the well being of poor Muslim weavers and laborers who relied on daily wages, and for whom curfews also constituted a form of economic repression. Physical violence was not as prolific or as drawn out in Varanasi as in other cities in north India. However, a profound sense of distinction was nonetheless (re)constructed between communities in the course of recurring curfews and enforced restrictions on mobility that these engendered (Brass 2006).

Informants suggested that episodes of local and national violent conflict, in tandem with the pernicious rhetorical marginalization of Muslims in the early 1990s, had influenced a broader process of distinction and distance between different religious groups which had reconfigured spaces of economic peace. This was made manifest in a decline in trust and respect, as well as the intentional distancing between respective business classes (see Assayag 2004; Froystad 2005). A respected elderly resident of Madanpura and current small-scale *gaddidār* sensed a tangible shift in the nature of intercommunity interactions following a series of riots in Varanasi in the 1970s. Before then, Mohammed Sadiq recalled the warm welcome he would receive as a *girhast* visiting Hindu *gaddidārs* at Chowk with a box of saris slung over his shoulder. During *Ramzān* (Islamic month of fasting) just as he was about to return home to break *rozā* (fast) it was very common for a Hindu *gaddidār* to stop him and insist that he break *rozā* at their place. However, today he observed a completely altered scenario. Rather than welcoming Sadiq as "*vyārpārī*" ("businessman") the Hindu traders were apparently more likely to utter: "*Are Miā ko todi jaghā hone do!*" ("Move away from this place, go from here!"). Sadiq recognized this transformation in behavior towards him, from respect to virtual dismissal, as the corollary of the economic squeeze. But he also placed it within of a broader process of sharpening differences between Hindus and Muslims in Varanasi since the early to mid-1990s as the Hindu right increased its influence in the state and national government.

Just as the form and mood of peace may shift through time, peace can also involve a variegated spatial and social reality. Even during periods when intercommunity relations were generally regarded as tense, the industry was perceived to represent a source of cohesion and cooperation. A Muslim *zarī* dealer from Madanpura described how:

> When the riots take place then the sari business is in need of money, and so this actually brings them together. In riots "peace committees" composed of the Hindu traders and Muslim weavers amongst others are set up to solve the problem or conflict in order to start trading again. The weaver and trader come to me asking me to help start the business again.

Similarly, a Muslim small-scale powerloom manufacturer and social worker from Madanpura recalled how Hindu traders at Chowk would keep him informed about when was a "good" (safe) moment for him to visit the market for trade with them. Sensing the potential for tension they would advise him to wait a few days, or send one of their party to collect saris from him. In this way, further difficulties were avoided and business could continue relatively unabated. Both cases demonstrated how the maintenance of everyday peace within a wider context of tension and immanent violence was in many ways a by-product of the imperative placed on the reproduction of intercommunity transactions and their businesses.

Particular spaces of intercommunity peace were constructed as "normal" vis-à-vis episodes of violence, which were "abnormal" or aberrations. During periods of Hindu–Muslim tension at local or national scales, relations in the silk sari industry were represented as harmonious compared to those "outside." Hindu *gaddidārs* spoke of protecting Muslim weavers, as one Hindu trader, Kishan Lal describes:

> If a riot takes place for example, the Muslim weavers will be safe here; there are four gates [his colleague later corrected that there were actually six gates and Kishan Lal admitted his error] into Kunj gali, so all the gates would be locked and whoever was inside would be safe. The Muslim weavers would be safe and sound, while outside here … now that would be a different matter.

By implying that Muslims "outside" Kunj gali in the majority Hindu *mohallā* were not guaranteed to be safe Kishan Lal acknowledged a real possibility of Hindu–Muslim conflict. By making this distinction between the relative security of spaces within the silk industry and without during such periods, Kishan Lal points to the fundamental bond between these communities, money, as well as the careful balance between brotherhood and enmity. At other times, however, Hindu–Muslim violence was regarded as characteristic of "other" cities and not typical of relations in Varanasi. In 2006, reports appeared in the local newspapers about Hindu–Muslim riots in Mau, a city 120 kilometers northeast of Varanasi which was also home to a significant Muslim artisan population. Muslim informants were adamant that such events would not inspire similar unrest in Varanasi because their city was the home of "Hindu–Muslim brotherhood." These exchanges reveal the ways that "peace talk" was spatially contingent, and differentially extended across different sites and scales within the market and the city as well as the surrounding region and nation. Within this fluid and potentially fragile peaceful landscape, Muslim weavers and traders sensitively negotiated intercommunity relations and their economic ambitions, as I explore in the final section of this chapter.

Economic Transformation and the Maintenance of Peace

In 2007 the visible intensity of sari showrooms and shops that lined the main street, and many of the lanes, in the majority-Muslim neighborhood of Madanpura was evidence of the upward mobility of some Muslim weavers into the business of silk trading. In the 1950s just four or five families from the *mohallā* exported their goods directly to firms outside the city; now over a hundred do so. Their business is conducted with buyers across India and also exported by firms based in Delhi and Mumbai to the Middle East, Europe and America.

With the growth of the market in Madanpura a particular air of confidence emanated from the trading classes who compared their market's success and service with that at Chowk. While relations between these two markets were generally cordial and cooperative, competition was regarded as endemic in the trade. As a successful Muslim businessman acknowledged, this could, at times, create animosity between the different religious communities: "There is rivalry. It's like in a family – if there are four daughters-in-law then there are fights. Similarly, there are fights in this industry. Both communities have the same business so it's obvious that there will be bad feeling towards each other." It was apparent, however, that as "daughters in law" they were not equal; rather, Muslim traders (and weavers) accepted their inferior place in the hierarchy as the more recent addition to the family.

During certain periods, competition between Madanpura and Chowk was perceived as particularly hostile, especially when business rivalry was constructed along religious lines. In 2002, the police and local media portrayed a murder incident that took place just outside Madanpura as having been committed inside the neighborhood and as religiously motivated, rather than a purely criminal act. A period of tension followed which led to a week-long curfew in the neighborhood. Hindu traders at Chowk sought to capitalize on the event for their own financial gains and led a malicious propaganda campaign discouraging buyers from going to the market at Madanpura. Notices were apparently circulated amongst businesses in Gujarat claiming that Madanpura was unsafe and that visiting traders were likely to be robbed. Members of the Madanpuria trading classes interpreted this as an explicit attack on their growing market, which was apparently challenging Chowk's established business.

The rhetorical violence committed against their reputation by competing Hindu traders acted to remind Muslim businessmen of their vulnerability as a minority religious community, particularly during periods of tension. The event also exposed how practices of economic coercion by the dominant Hindu community played a role in structuring intercommunity relations, and also, concurrently, the future practices and orientations of Muslim businessmen. The upshot of these subtle as well as more overt economic pressures was that Muslim traders were constantly forced to draw upon their powers of resilience and to think creatively about how to safely expand their businesses. For instance, some informants talked about actively nurturing business connections outside of Varanasi and Uttar Pradesh so as not to antagonize the "other" community in "their" established territory, while others emphasized the need to develop relations with the local politicians in the hope that their interests would be protected.

Five years after this incident and the dominant narrative in both Chowk and in Madanpura was that relations between the two markets now centered on cooperation. Varanasi Madanpura Sari Dealers Association was set up in 2005 in response to a demand for coordinated business and organizational thinking within an increasingly independent Muslim market-sphere. The association also provided an official channel of communication with *Banaras Udyog Sangh* (Banaras Enterprise Association) which was based at Chowk; it emerged in conjunction with an apparent increase in cooperation between the markets. Muslim and Hindu traders both spoke of the common practice for the markets to "speak to each other" for instance, if a particularly dishonest customer dealt with Madanpura and then proceeded to Chowk, businessmen in Madanpura would warn parties at Chowk ahead of the visit to be wary and vice versa.

The general consensus was that lessons had been learned following the negative impacts experienced by businesses during periods of religious based conflict and curfews. Furthermore, it was frequently remarked that Hindu–Muslim riots were the product of politicians' ambitions rather than the desires of ordinary people (*ām ādmi*). And amongst Varanasi's "*ām ādmi*" there was an eagerness that they would not be duped into acting divisively in the future. Within Madanpura's more affluent business classes the outward impression that was self-consciously projected was one of actively fostering intercommunity cooperation. This cooperation was widely acknowledged as critical in sustaining their own businesses, as well as the livelihoods of Muslim weavers within their neighborhood, for whom they articulated an element of responsibility. But while members of the Muslim business community continued to construct the notion of Hindu–Muslim accord, they were pragmatic about their position within this dynamic, and the potential risks as well as constraints for a community that was so dependent upon Hindu traders. There was a collective desire to be independent from the Hindu community, particularly in terms of distancing themselves from business at Chowk. However, this was not wished at the expense of peaceful intercommunity relations. The widely acknowledged reality was that as Muslims working and living within a Hindu-dominated economic, cultural and political milieu their hopes and aspirations were intimately connected with the nature of intercommunity relations and the reproduction of everyday peace. This serves to highlight the differing degrees of peace and the ways in which contrasting landscapes of peace are actively negotiated. As Darling (2014) has argued, peace can be mundane in its appearance and needs to be sensed through the routine state of things. In understanding everyday peace we therefore need to be attentive to both the visible and conscientious forms of peace agency, which find expression in the immediate face of tension and conflict, as well as the less visible articulation of constructive agency over time.

Conclusion

This chapter has documented the role of the economy as a central site within which everyday peace between Hindus and Muslims was produced and reproduced in Varanasi. My analysis extends a theme introduced earlier in the book around forms of "peace talk" and the imaginative potential of everyday coexistence, to actually illustrate the uneven and shifting realities of interactions in the silk sari market. The chapter shows that discourse and action interact in geographically situated ways in the spaces of the silk market to construct and constitute everyday peace. And, it advances our understanding of peace as political, challenging the idea that peace is a condition that emerges only once politics has been done and justice achieved.

It also reveals the political undercurrents of "peace talk." Just as Richmond (2005) has sought to critique the normative assumptions that surround the notion of "liberal peace" in international peacebuilding initiatives, the same questions need to be asked of local narratives of peace. In whose image is the notion of everyday peace constructed, whose interest(s) does it serve, and what practices of power does it conceal? The notion of "Hindu–Muslim brotherhood" reflected a particular intercommunity reality based on shared economic fate, which also functioned as an imagined geography of peace even where experiences of intercommunity exchange, trust, affection and loyalty were elusive. But, as a normalizing narrative I argue that "Hindu–Muslim brotherhood" served to veil or depoliticize inherently unequal practices of power within these real and imagined intercommunity economic realities. Within a Hindu-dominated economic and political milieu, coercion and consent often bound Muslims into economic relationships with more powerful Hindu traders. Indeed, given the intricate and interdependent web of Hindu–Muslim relations that constituted the silk industry, everyday peace was a necessary by-product of the industry's continued success. The normative discourse of intercommunity harmony therefore acted to forestall questions and consequently defer tensions around the iniquitous conditions experienced by Muslim weavers that were integral to the reproduction of economic peace. The conclusion is that uneven power relations in fact often underpin peace; that practices of inequality and injustice are constitutive of everyday peace rather than resolved or realized through intimate, daily encounters.

In Chapter Four I argued that residents often accepted experiences of inequality and injustice or sought alternative avenues to minimize patterns of discrimination in order to sustain everyday peace. This chapter adds another dimension to that argument and critique of Galtung's conception of "positive peace" by showing that the reproduction of everyday peace can actually *depend* on the perpetuation of inequalities and injustice. In these economic spaces, conspicuous efforts to gain equality and justice by Muslim participants would undoubtedly have compromised peace. So, while it could be reasonably argued that all actors within the industry had a stake in the reproduction of economic peace, in the context of a Hindu-dominant milieu that differentially constructed the city's Muslims as a threat, the responsibility for reproducing peace appeared to rest more substantially on the shoulders of Muslims (see also Heitmeyer 2009). Consequently they advanced their economic ambitions through longer-term strategies grounded in pragmatism, acceptance and resilience. Emergent Muslim entrepreneurs were strikingly cautious about disrupting the economic balance between themselves and the Hindu trading classes in order to protect their economic ambitions. During my fieldwork, no Muslim traders owned a shop in the main market at Kunj gali, which is both indicative of the spaces apparently unavailable to Muslim businessmen and their preference for forging new networks independent of the established Hindu-dominated markets.

The experience documented here of everyday economic peace was ultimately contingent on maintaining distinctions, reproducing boundaries and knowing one's place (see Froerer 2007 for a similar argument around caste). This study therefore adds nuance to arguments that celebrate urban micropublics as having the potential to destabilize and transform ideas of difference (see Amin 2002). I argue that, whilst this can be the case, we need to pay greater attention to the ongoing patterns of social and material inequality that underscore notions of difference and impede truly transformative relations. The chapter documents complex and shifting encounters across difference that resemble both connections for some, but also sustained disconnections for others, which contributes to maintaining everyday peace but also embodies forms of fragility and potential fracture.

Thus far, this book has documented dimensions of everyday peace through different public spaces and events to show how practices and experiences of peace are closely connected to citizenship realities for Varanasi's Muslim residents. In a context where citizenship for India's Muslims has often been partial or compromised, it is interesting that initiatives to create alternative spaces for citizenship are intimately related to ensuring that everyday peace is sustained. An important strategy for achieving this has entailed a struggle for inclusion based on keeping a low profile, not provoking reactions from society or the state and appealing to secular narratives. The final empirical chapter of this book maintains a focus on public space, but it presents a rather different scenario as it documents a particular event, the arrest of an Islamic religious teacher, to examine what happened when local Muslim residents chose to become visible, and publicly challenge the status quo.

Chapter Seven
Becoming Visible: Citizenship, Everyday Peace and the Limits of Injustice

Introduction

The preceding chapters have documented some of the ways in which Madanpura's Muslim residents characteristically worked to maintain everyday peace where events threatened to unravel the routine order of things. In spite of the uneven practices of citizenship that were reproduced alongside peace, maintaining peace was of paramount concern, especially where intercommunity violence and associated curfews would have resulted in far greater injustices for them. Consistently denied full citizenship within different city publics, Muslim *Ansāris* typically demonstrated degrees of acceptance. Instead of publicly contesting injustices and overtly voicing their grievances, they subtly and pragmatically constructed alternative spaces of citizenship or "counter-publics." Such strategies enabled them to keep a low profile in the city, to avoid confrontation and conflict, and yet at the same time continue to engage with the state rather than rejecting it outright. In this chapter I focus on a more arresting and visible manifestation of Muslim

Everyday Peace?: Politics, Citizenship and Muslim Lives in India, First Edition. Philippa Williams.
© 2015 John Wiley & Sons, Ltd. Published 2015 by John Wiley & Sons, Ltd.

claims to citizenship, through which residents actively sought to question injustice and reconfigure citizenship rights in order to overturn dominant narratives within everyday relational space.

After introducing the circumstances of the *Maulānā's* arrest the first section of the chapter documents how these events provoked questions about citizenship as founded on contrasting "solidaristic, agonistic, and alienating strategies and technologies [that] constitute ways of being political" (Isin 2002, p. 29), and it highlights the relationship between visibility, community and citizenship. Second, I explore how overlapping solidarities cohered around notions of injustice, recognition and rights to the city which enabled the possibility of collective protest, and the production of activist citizens. Third, I examine how the protest, as an "act of citizenship," was dependent upon its potential answerability, which required it to be regarded as legitimate and civil and therefore constitutive of everyday peace. Finally, if "citizenship" is to be conceived of as an act that demands a multilevel response, I propose that greater attention should be directed towards the role played by contingency and context in shaping the potential for an answer, and therefore for an "act of citizenship" to be actually realized and everyday peace sustained. Against the wider backdrop of the book, this particular event proves especially illuminating in demonstrating the accepted limits of injustice and inequality when it comes to reproducing everyday peace in the margins.

Rajasthan's police arrested Maulana Abdul Mateen in the lanes of Madanpura as he was walking home from work with a colleague. The police accused the Islamic teacher of being implicated in synchronized bomb blasts which had taken place in Jaipur two months earlier, and for which the Indian-based Islamist organization, the Indian Mujhaideen, had publicly claimed responsibility. After Abdul Mateen was driven away, his colleague immediately alerted local residents about the events. Family and colleagues of the *Maulānā* wasted no time in appealing to the local police to release information about not only his location but also the indictments levied against him. In spite of the persistent requests made through these formal channels of enquiry, their efforts proved futile and the important information failed to materialize. Lacking concrete information about the *Maulānā's* circumstances the primary concern was that like other Muslim men, whose stories appeared in the newspapers from time to time, he would disappear or be indefinitely detained by the police without an official charge sheet filed against him. Motivated by frustration with the administration and a degree of fear and uncertainty, his family and local personalities promptly mobilized residents in the neighborhood to coordinate a mass public protest asserting the *Maulānā's* innocence and calling for his release. So, Muslim residents came on to the streets in their thousands and staged a protest until the second night after the *Maulānā*'s arrest when Rajashtan's police administration made the rather surprise announcement that the *Maulānā* would be returned to Varanasi. The protesters' demands had been answered.

The Protest as an "Act of Citizenship"

As discussed earlier in the book, citizenship has increasingly been seen not merely as a legal category, but as a set of discourses and practices that are translated unevenly across unequal social groups and local contexts (Lister 1997; Werbner and Yuval-Davis 1999; Chatterjee 2004; Benhabib 2007; Holston 2008). Scholars have shown how citizenship may also become articulated as a hegemonic strategy, which works to define particular groups or localities, to fix the power differentials between them, and then to naturalize these operations (Isin 2002; Secor 2004). Within the Indian citizenship framework, I argue that notions of the Muslim "victim" and "terrorist" have found widespread circulation. But, as Secor (2004) has shown in her research amongst Kurdish migrants in Istanbul, Turkey, these hegemonic strategies are never completely successful. By focusing on this public protest I show how Muslim residents came out on to the streets to publicly dispute the police administration's judgment and strategically position themselves in relation to others in the city to stake their claim to being rightful and deserving Indian citizens. This particular act of protest demands attention because, as a break in the hegemonic order of things, it was intended to publicly recast dominant images of Muslims differently. The question is, to what extent did the protest succeed in doing so, and with what, if any, longer-lasting effect?

The protest performed by Muslim residents in response to the *Maulānā's* arrest may be interpreted as an "act of citizenship" through which actors constituted themselves (and others) as subjects of rights (Isin 2009, p. 371). Acts of citizenship are understood to represent a rupture in the everyday and explicitly attempt to overturn social-historical patterns of power through lived experience (Isin and Nielsen 2008). The struggle for citizenship is "not just about legal status, but for recognition as someone with an audible and corporeal presence that can be described as 'political'" (Nyers 2007, p. 3). Isin makes a distinction between "being political," as that of being implicated in strategies and technologies of citizenship as "Otherness," and "becoming political" as "that moment when the naturalness of the dominant virtues is called into question and their arbitrariness revealed (2002, p. 276)." "Acts of citizenship" are precisely about "becoming political." They represent "actively answerable events that anticipate a response," that is both "particular and universal" (Isin and Nielsen 2008, p. 10). It is through this dialogical process, of acting and reacting with others, that citizens negotiate their relative position with others (Isin 2008, p. 38). And as political moments it is through acts of citizenship that individuals or collectives may constitute themselves differently from the dominant images given to them (Isin 2002, p. 33).

By thinking about the city (Isin 2002) or even society more generally (Staeheli 2005) as a "difference machine," Isin privileges the notion of alterity to capture the immanence of "others" rather than the notion of exclusion which conceives of "others" as more distant or exterior. "The logics of alterity assume overlapping, fluid, contingent, dynamic, and reversible boundaries and positions, where agents engage in solidaristic strategies such as recognition and affiliation, agonistic strategies such as domination and authorization, or alienating strategies such as disbarment across various positions *within* social space" (Isin 2002, p. 30).

Political geographers have uncovered some of the ways in which citizenship and space are closely entangled (Smith 1989; Dowler and Sharp 2001; Secor 2004; Miraftab and Wills 2005; Pain and Smith 2008; Dikeç 2009). Secor (2004) draws on de Certeau (1984) and Lefebvre (1995) to view citizenship as a technique of spatial organization or "strategy" through which individuals can make claims to the city. She shows how Kurdish women in Istanbul negotiate and differentially assert their citizenship by deploying contrasting tactics of anonymity and strategies of identity within and across different urban publics. This enables them to actively hide and unhide their cultural identity as Kurds and/or as Turkish citizens, at once constituting and contesting political moments, in their everyday lives. Understanding how political identities may be reconfigured in relation to space (and time) opens up the idea of citizenship to alternative possibilities, such as those of multiple public spheres (Fraser 1992) or multilevel citizenship (Yuval-Davis 1991).

The strategy of hiding and unhiding particular identities and affiliations also speaks to the complex, fluid and contested relationship between citizenship and community (Staeheli 2005). In theory, India's Muslims are recognized as rights-bearing individuals within the Indian liberal polity. However, the constitutional framework requires that in practice, they have to lay claim to their citizenship via the notion of their minority community, as Muslims. Yet, as a minority religious community which has experienced patterns of material and rhetorical discrimination, the possibility of asserting rights to citizenship via the language of "Muslim" rights has proved highly contentious. Alongside Chapter Four, this chapter shows how Madanpura's residents were sensitive to the ways in which their individual and "community" image were entangled and constructed within different city and national spaces, and accordingly to the differential risks that publicly asserting their citizenship and minority identity entailed. Citizenship practices were therefore strategically articulated. Of course, however manifested, "communities" are rarely homogenous entities but are instead internally fractured by an array of contradictory identities and subjectivities (Staeheli 2008). A "community" may become visible in public space where degrees of commonality and solidarity exist between its members; these may emerge organically or be orchestrated by different actors. As Secor (2004) and Staeheli (2008) have argued, visibility may be an important strategy for empowering communities and ushering them into the polity and citizenship (also see Domosh 1998; Dwyer 2000; Marston 2002), but at other times it may undermine the partial practices of citizenship that are realized. The relationship between visibility, community and citizenship is therefore a contingent and shifting one, which is strategically negotiated within different urban publics.

Solidarities: Injustice, Recognition and Rights to the City

"By theorizing acts or attempting to constitute acts as an object of analysis, we must focus on rupture rather than order but a rupture that enables the actor (that the act creates) to create a scene rather than follow a script" (Isin 2008, p. 379). The rupture, in this case the arrest, generated notions of injustice which prompted local residents to mobilize and create the "scene" by appealing to overlapping solidarities which contested injustices and concerned demands for recognition and rights to the city. Solidarities speak to the capacity to identify with others and to act in unity with them in making claims for justice and recognition. Through the protest, a figurative sense of unity was forged, not only with the *Maulānā*, but also between members of the neighborhood who were brought together by the broader security concerns that the arrest engendered, and in so doing constituted shared space. Three solidaristic acts or narratives may be identified. Firstly, there was a unified confidence within the neighborhood that the *Maulānā* was innocent and therefore unjustly accused by the Rajasthan police administration. Secondly, there existed a widespread concern that given the failure of formal lines of enquiry to locate the *Maulānā* he would personally endure a fate similar to that of other innocent Muslims involving long-term detention without trial or even the lack of potential for a fair trial. This was underpinned by a widespread distrust by local Muslims of the police, whose actions were believed to be universally prejudiced and whose agenda was to protect Hindu majoritarian interests (see Chapter Five). Thirdly, residents saw their own vulnerability through the *Maulānā's* fate. The arrest of the

Maulānā had materialized anxieties that many already experienced about being visibly Muslim in public space. The urgency to see the *Maulānā's* arrest revoked was therefore also about restoring the reputation of the neighborhood at large, which presently risked being portrayed as a universal site of possible Muslim terrorists.

These acts of solidarity spoke of and embodied constituent claims to rights which primarily concerned matters of justice, or injustice that were felt at different scales and sites, notably vis-à-vis the state, but also within society and the public spaces of the city. As Isin points out, the substance of citizenship may be "rights," but rights are not substances, rather they are relationships which reflect dominant sites and actors of citizenship (2009, p. 376). As a unique moment in which local actors "claim the right to claim rights" (Nielsen 2008 after Arendt 1986), this site proves particularly revealing of the dominant forms of situated power that structure Muslim experience as relational in Varanasi, and in India more widely.

Claims to rights were made in three contrasting ways. First, rights were interpreted with respect to notions of relational justice. As is more often the case, ideas about justice find clarity in the face of acts of injustice. And for Muslim *Ansāris*, notions of injustice were routinely understood in relational terms vis-à-vis the Hindu majority. During my fieldwork period this conception of injustice was demonstrable in the shape of activities going on in Jammu and Kashmir at the time around the issue of the Amarnath Shrine, and the tactics used by the state to disperse protests in different parts of the state. Jamaal bhai, a Muslim sari businessman in his mid-thirties voiced his frustration:

> In Jammu, Hindus were protesting peacefully and the administration fires *water* on them. Meanwhile in neighboring Kashmir when Muslims were also protesting peacefully the administration fired [bullets] on them.

The recurring mention of this incident by residents in Madanpura reflected the common narratives that were circulating around the neighborhood as residents sought to make sense of the events amongst themselves. The state responses to ongoing incidences in Jammu and Kashmir captured local imaginations since the evidently differential treatment meted out to Hindus and Muslims in seemingly proportional circumstances represented a classic example of the general state bias in Indian everyday life. A more prosaic experience of injustice was illustrated in the course of my conversation with two middle-aged Muslim men in a *chāy* shop, just a few months after the *Maulānā's* arrest. One of the men called to mind his everyday experiences of being served at his local city bank. He expressed his disgruntlement that

Muslim customers consistently received inferior service compared with their Hindu counterparts:

> If I went in the bank in this outfit [stereotypically "Muslim" dress] to pay in a check and the check was supposed to be placed at the top of the pile, instead it would be placed under ten other checks and so I would be served last ... On one occasion while at the bank I noticed that another Muslim, who was dressed up in shirt and trousers, had come to collect his check but since the clerk couldn't figure out whether he was a Muslim or not, his check was cashed immediately, while myself and other Muslims had to wait.

Ideas around justice were therefore understood not in terms of retribution or revenge, but rather about when it is fair to be treated the same, and when it is fair to be treated differently (see Kabeer 2008, p. 3). Under India's secular constitution, Muslim residents expected that whatever their religious affiliation the state should act fairly and impartially towards its citizens (see Chapter Four). The narratives imparted on this matter frequently referred to local state and societal practices that undermined this concept of justice and fairness.

The second claim was to the right to equal recognition. This was closely associated with demands for relational justice. Typically, demands for recognition insist on both recognition of the intrinsic worth of all human beings as well as the recognition of, and respect for their differences. Calls for the *Maulānā's* release were expressed through a language that primarily emphasized his everyday civility and good, caring character. The *Maulānā* was repeatedly referred to as a "noble" and "helpful" man, while his charitable organization was constantly evoked as evidence for this. It was anticipated that such a commitment to benevolent activities marked the *Maulānā* out as a deserving Indian citizen. In calling for their recognition as good Indian citizens, Muslim residents wished to be seen as equal to Hindus, and like them, should not have to experience the risk of being picked up off the streets by the authorities, accused of terrorism, just because they were Muslim. Residents regularly asserted their membership within the nation, stating that "we are all Indian citizens, Hindu, Muslim, Sikh and Christian." The challenge presented to Madanpura's Muslims was that everyone did not appear equal in reality. Even whilst informants asserted that "[w]hat's right is right for everyone, what's wrong is wrong for everyone" they were painfully aware that this was not the case. The question of differential recognition also arose over the matter of what is seen by the Indian state to constitute terrorism in the first place, and what is not.

The VHP [right-wing Hindu group with links to the BJP] are openly doing violence [for example bomb blasts and in instigating riots] but they are not called terrorists. But the community that is cooperating [i.e. Muslims] – is the one that's being accused of enacting terrorism. Even while the VHP and BJP are openly carrying out violent acts in the country the administration are simply holding up their hands and condoning it.

For Muslim residents more generally, the arrest locally reaffirmed the state's incomplete recognition of them as equal citizens of India. It was clearly understood that their actions would always be interpreted as intentionally subversive whilst comparable violence by Hindus would not be subject to equivalent attention or concern. In the act of protest, Muslim residents had something to say about themselves as citizens; by evoking notions of civility and the good citizen they sought to upturn dominant constructions of themselves as Muslim terrorists or even as passive victims. Yet, the demand that they be recognized as the same, and treated the same nonetheless publicly (re)constructed Madanpura's Muslim *Ansāris* as a visible "community" at once different to the city's majority Hindu population.

The third constituent claim concerned Muslims' right to the city, and the possibility of occupying and interacting within public spaces free from suspicion. The tension was that the *Maulānā's* arrest had yet further compromised notions of security in public spaces, as Farhan Sayyed, a Muslim social worker and ex-ward corporator explained:

In Madanpura people who have a beard and dress in a particularly Muslim way are advised to avoid going to crowded places like Dashaswamedh ghāt and other crowded places because they will be harassed by the police... this is a recent thing... People are feeling "insecure" in that [Muslim] look, they are worried about what might happen to them... they could be arrested or something.

Intimately related to this question of freedom to move within urban spaces was the matter of visibility; in particular, the perceived risks involved in being marked out as a Muslim in majority Hindu space (see Cresswell 1996). Nevertheless, there was widespread disquiet amongst residents that the *Maulānā's* arrest during a period of heightened concern about local "Islamist terrorism" had further compromised the potential for them to be treated with equal respect in public spaces.

The demand for their right to the city was made through a direct engagement with urban space: the protest physically appropriated the city's streets. It is worth highlighting that the protest was conducted within the boundaries of Madanpura. But even whilst it was constituted within notably

"Muslim space" the presence of thousands of bodies did serve to block two of Varanasi's main thoroughfares at Madanpura, Reori Talab and Kamachcha, and as such, represented a disruption to the circuits of everyday city life more generally. Friends and informants living outside of the neighborhood recounted how during the course of the protests their movement across the city had been severely curtailed, causing them great inconvenience. Brahmin Hindu friends also vocalized the apprehension they had experienced as they perceived the potential for the protest to turn violent, and for riots to erupt in the city. Fearing the possibility of being caught up in such events, many had actively avoided visiting localities adjoining Madanpura during the protests and the days afterwards. News of the protests initially travelled by word of mouth between family members and friends and then through the local Hindi newspapers. Although not intentionally inflammatory in their reporting, local newspapers carried photos that depicted apparent mobs of angry Muslims on the streets being held back by the police. Given Madanpura's centrality within Varanasi's imagined geographies of security and insecurity, there was a tacit understanding that events within Madanpura could have more far-reaching implications for everyday peaceful relations within the city at large.

The role of space was central to both the manifestation of the protest and the possibility of a rejoinder by the police and wider population in Varanasi. By conceiving of space as a configuration, it is "never simply a passive background of becoming political" (Isin 2002). Here, the neighborhood was both disrupted, and, as the product of interrelations, it also proved to be the source of that disruption. Space, as relational, therefore constituted a transformative potential for Madanpura's residents and the claims embodied in their act of protest (Dikeç 2005, p. 181). However, the legacy and durability of this transformative potential was nonetheless circumscribed within Varanasi's situated geographies.

The Act as Answerable

It is very important that we succeeded against this injustice because today they may arrest one person and by tomorrow it might be another. And for each and every blast in future they would say it was because of someone in Madanpura.

As this quote by Aijaz, a middle aged, middle class Muslim sari agent implies, at the heart of the protest was an attempt to redress issues of perceived civic injustice that had been committed, not just against the *Maulānā's* personal freedom, but also more generally against Madanpura's residents and their neighborhood's reputation. But to succeed against "this injustice" demanded eliciting a particular response from the police, and wider Varanasi public caught up within the scene.

By acting in public space and disrupting the order of everyday city life, the protest brought together a range of actors who became involved in the scene, both actively, as well as by chance. Prabhudatt, a close Brahmin Hindu friend remarked how, "[t]here was so much tension, it was like there was a riot in the city – but a controlled riot." Although he lived in a neighborhood three miles south of Madanpura, the tension that surrounded the protest meant that Prabhudatt had nonetheless found himself drawn into its politics, as he took views on the questions of justice that it raised and the position of India's Muslims more generally.

Isin draws on Levinas (1978) to theorize the question of political acts and justice. He maintains that an act always assembles a spectrum of others beyond those actors initially caught in the scene that the act creates. This assemblage of actors, gathered together through space, will provoke a question of comparison, coexistence, visibility, difference and all that stands for justice. Moreover, precisely because justice is "an incessant correction of the symmetry of proximity," seeking justice will involve answerability. In summary, "[j]ustice seeks answerability when one makes a claim upon the other in the presence of another" (Isin 2008, p. 36).

The notion of answerability is double-sided in that it speaks to both the universal and the particular (see Bakhtin 1993). Universal obligations of the act of citizenship focus on the immediate demand of the present, as well as the potential of other possibilities in the name of justice and democracy-to-come. Meanwhile, the act also addresses the particular, where subjects should give an account of their own exceptional uniqueness. The act of citizenship therefore not only speaks to the other, but also to the self (see Isin 2008; Hsu 2008, p. 249).

Where the quest for justice necessarily involves seeking answerability, orientating oneself with responsibility towards others becomes essential (Isin and Nielsen 2008, p. 31). In the hope of securing the *Maulānā's* release, the claim was made that the *Maulānā* was an innocent and deserving Indian citizen, and by association Madanpura's residents more widely were civil and good. Moreover, it was paramount that the protest itself embodied this degree of civility towards others. Had the protest descended into descent or dissident practices, the potential for the protest to be answered effectively and for justice to be achieved would have been thoroughly undermined. Had the protest broken down – dissolved, collapsed or fragmented into "uncivil violent acts" – then it would not have constituted an act of citizenship (Isin 2008, pp. 10–11). In maintaining the answerability of the act, local neighborhood residents, as well as political and religious leaders, actively worked to keep the protest civil and good-natured as people assembled in the streets over two days.

Two local residents, Farhan and Nadeem bhai, described how they took active roles in keeping the protest civil by mediating between the protestors and the police on the ground. Both of these men were in their mid-thirties and involved, to differing extents, with social work in the neighborhood. They held progressive views about the kind of direction business as well as education should take if the standard of living in their neighborhood were to improve. Farhan and Nadeem bhai pointedly objected to the dominant narratives imparted in everyday city conversation that constructed Madanpura as a backward locality, inhabited by dangerous and uncivil Muslims. Such images jarred sharply with Madanpura's actual lived realities as well as inhabitants' self-representations. The need to keep things civil was of paramount importance not only in the context of this immediate injustice, but also with respect to more long-term relations with the police administration and wider Varanasi society. Farhan was explicit on this matter, perceiving that without the work of mediators such as himself there was a strong potential for the administration to create problems for their neighborhood in the future.

Both men recognized that there were risks involved in facilitating between the police and local residents. Whilst attempting to manage and calm protestors within the crowd, Nadeem bhai was photographed by a newspaper journalist. The photo was published the next day in the Dainik Jagran and apparently represented Nadeem bhai as an angry protestor, struggling against the police.

> That photo was very critical – it looked as if I was one of the protestors, the way my arm was raised in the air. But in fact I was protecting the Senior Police of City. I was trying to make sure that the crowd didn't get so out of control, otherwise the police might have reacted and fired bullets at them... After seeing this photo I took it to show him [the Senior Police of City] how I was being portrayed because I was worried that people might see this and think I was a protestor out to cause trouble. But he said he knew the reality of the situation and that I shouldn't worry.

Nadeem bhai recalled how his neighbors had warned him against getting involved with the administration and intervening in such hazardous business. They believed that he risked becoming vulncrable to arrest, especially when condemning photos were published in the press. Farhan was also careful not to betray the trust of his fellow residents. Whilst evidently proud of the praise he had received from the police, who had subsequently recognized his assistance in peacefully managing the protest, he resolutely refused to impart critical information to the police about the identities of key protagonists. Farhan was anxious that communicating such information would inevitably place him in danger of reprisal by local residents and earn him the derogatory label of being a *mukbhir* (informant). This would not only undermine his local social and working relations, but could also inhibit the potential for future brokering opportunities within or between his neighborhood and the administration. The work of Farhan reveals the important role played by local actors in mediating risks and disputes within the protest in order to ensure that it constituted a fertile context for answerability. Such mediators occupied a more ambiguous position with respect to the act of citizenship, at once becoming activist citizens through the protest, they also sought to structure the scene, and shape the possibility for a particular kind of response.

Yet, the possibility of the protest being recognized as responsible and answered, also depended to an extent on more contingent factors, something that Isin (2002, 2008) does not fully explore.

Answerability and Contingent Geographies of Citizenship

Developing this point further, I suggest that the answerability of this act of citizenship depended not only on the protest remaining civil, it also hinged upon a range of interacting, situated and contingent geographies. Things could very easily have been different (Shapiro and Bedi 2007, p. 1), had events cohered differently in place at that time. As an approach, and not a theory or strategy, feminist geopolitics insists upon the contingency of place, people, and context in making change. Here, I employ contingency to refer to a conjuncture of events without perceptible design. Whilst it is possible to extend the notion of contingency indefinitely, I wish to highlight some of the immediate and interacting conjunctures that were central to the realization of the protest. These pertain firstly to the chance of the protest materializing at all, and thereby creating a rupture in the everyday order of things, and secondly to the possibility that demands made by the protestors were actually listened to and effectively answered.

Concerning the first of these situated contingencies, I suggest that the possibility for a rupture in the usual order of things, which required collective action, was conditional on Maulana Abdul Mateen's social context. As I have discussed earlier, during my period of fieldwork between 2006 and 2010, it was not uncommon to read stories in the newspapers about Muslims, usually males, who had been arrested by the police and taken away with little or no explanation. A month after the *Maulānā's* arrest, a small-business owner from Lucknow, Shahbaz Ahmed, was also arrested in association with the Jaipur bomb blasts, this time in a joint operation between the Rajasthan and UP police administration. As in the case of the *Maulānā*, popular opinion believed Shabhaz Ahmed to be innocent, but his arrest did not provoke a local outcry. At least 83 days after Ahmed's arrest he remained in custody, with a charge sheet yet to be filed for his case (*Times of India*, November 16, 2008). It is reasonable to assume that the context of the arrest in Lucknow embodied a potential for geographies of solidarity to emerge comparable to those witnessed in Varanasi, but, curiously, they did not. This discrepancy might be explained in light of the contrasting personal contexts and respective relational settings of those accused.

Maulana Abdul Mateen was a respected teacher at Jamia Salfia, Varanasi's Arabic University, founded by Madanpura's Ahl-i-hadith members. His position at the university inevitably afforded the *Maulānā* an extensive close network of colleagues and students, as well as Ahl-i-hadith followers from the neighborhood and wider city. Ahl-i-hadiths are well represented in the merchant classes of Madanpura's Muslims, and have traditionally comprised the neighborhood's more affluent and well-connected residents including one time city mayor, Mohammed Swaleh Ansari. The *Maulānā* belonged to one of Madanpura's most venerable and prosperous silk-trading families and was thereby a part of Varanasi's longer history. He was evidently positioned within strong community networks, which lent him close links to actors capable of leveraging wider support amongst different groups in the community and beyond.

Conversely in Lucknow, Shabhaz Ahmed was not similarly endowed with a network of support at the time of his arrest. A relative newcomer to Lucknow, Shabhaz Ahmed had moved from Bhadohi (also in UP, west of Varanasi) in 2004. Leaving his father's carpet industry and the security of local community networks he arrived in Lucknow to set up his own business. At the time of his arrest, Ahmed was running a career consultancy firm, but as an independent business entrepreneur his community ties were relatively weak and less extensive than the *Maulānā*'s in Varanasi. Lacking the social standing and connections with influential and extensive networks, Shabhaz Ahmed did not receive immediate popular support in opposition to his arrest. Whilst a subsequent fact finding report protesting Shabhaz Ahmed's innocence was compiled by the People's Union for Civil Liberties (PUCL) and People's Union for Human Rights (PUHR) and widely publicized in October 2008, it did not exert adequate pressure on the administration for it to reconsider his arrest (see PUCL 2008).

That the protest was actually listened to and positively answered by the police administration points to a second set of contingent realities. It was not unheard of for the Indian police administration to respond with force against protests, as was seen in the case of Nandigram in West Bengal in 2008, and against Muslims in Kashmir as mentioned earlier. The question is why were the demands of protestors actually met in this particular circumstance? It is important to highlight and emphasize that the police administration of Rajasthan, and not of Uttar Pradesh, instigated the *Maulānā's* arrest. This enabled Varanasi's police administration, with the support of the Uttar Pradesh force, to play the role of mediator between the protestors and Rajasthan's authorities, without themselves being directly subject to blame by city residents. Different actors associated with the protest recognized that there was the potential for the local administration to work in the interests of the *Maulānā* and Madanpura's residents, rather than against them, in this circumstance.

The cooperative and moderate approach of the local administration was importantly shaped by the immediate events in the city context. The protests were coincident with the Sawan festival, which had brought hundreds of Hindu pilgrims into the city, and had already placed the administration under additional pressure.[1] The proximity of "Hindu pilgrims" and "protesting Muslim residents," crudely expressed but nonetheless imagined as such, embodied a perceived potential for conflict. The police wished to avoid this confrontation in the city and therefore worked hard to deliver an agreeable response. In a media interview, the Senior Superintendent of Police described how "the strength of feeling" demonstrated by residents in Madanpura had played a key role in influencing the UP administration's very "unusual" decision to intervene in the activities of another state's police administration (*Dainik Jagran*, July 28, 2008).

Thirdly and in relation, actions taken by local personalities were crucial to the act of protest not only being conducted in a civil manner, as I have discussed, but actually translating into a tangible and positive outcome. When the initial pleas by the Superintendent of Police (SP) and the Chief Minister (CM) to protestors to cease demonstrations went unheeded, the administration was forced to convene urgent meetings in the neighborhood. Local residents and leaders played critical roles in mediating between the protestors across different scales and sites. Public personalities such as the Mufti-e-Banaras and the Samajwadi Party MLA for Varanasi North – a member of the Muslim *Ansāri birādarī* – were drafted in to represent the Muslim community, and were invited by the police administration to discuss possible resolutions to the protests. As I have shown in Chapter Three, the Mufti-e-Banaras had secured a respected reputation for successfully negotiating moments of tension between different religious groups in the city. Moreover, his public dedication to the local Muslim population encouraged the protestors to trust that their interests were being faithfully articulated within senior levels, and to heed his calls that, although justified, the protest should not jeopardize everyday peace in the city.

Conclusions

This chapter has documented how negotiating citizenship in the margins, characterized by both material inequalities and histories of violence, is contingent on the possibility of reproducing everyday peace. It shows how, as an "act of citizenship," a protest created a rupture in the fabric of everyday city life, and, as an event, brought to light possibilities for Varanasi's Muslims that were invisible or unthinkable in the everyday, as the preceding chapters have discussed. Its focus on a particular "event" in the context of everyday peace therefore provides a productive opportunity to witness the boundaries of everyday injustice and narrative violence, whilst recognizing the limits of this particular moment and its longer-lasting legacy. The *Maulānā's* arrest and the public cessation of *his* citizenship demanded a visible act of citizenship by Madanpura's residents as they struggled to restore his freedom, *and* the neighborhood's reputation. In this moment, Madanpura's typically taciturn residents "became political," activist citizens. They actively constituted space and questioned injustice in efforts to overturn the circulating image of the "Muslim terrorist." Interestingly, this citizenship practice represented a strategic unhiding of Varanasi's Muslims. As Madanpura's residents congregated on the streets they represented a visible Muslim "community," both to themselves and the wider city public.

This event thus stimulates some wider implications concerning the conceptualization and realization of everyday peace, citizenship and justice for India's Muslims. First, it contributes towards understandings of peace as political, which is constituted through the ongoing negotiation, rather than reconciliation, of difference and inequality. Earlier chapters have shown how India's Muslims typically recognized acts of injustice and discrimination on the basis of their religious identity and numerical minority, and that rather than contest such injustices local residents expressed forms of everyday acceptance. Yet, the protest reveals how there were limits to acceptance, and limits to practices of everyday injustice where they posed explicit challenges to inclusion. By publicly contesting dominant images and provoking a necessary answer from different actors, the protest involved an element of risk that was deemed worth taking. The protest was regarded as the last option for securing not only the *Maulānā's* freedom, but also the potential for Muslims more generally to realize justice in future.

Struggles to minimize injustice were therefore carefully articulated in different relational and scalar contexts.

The way that this struggle for justice and citizenship was articulated speaks to questions of agency, a central theme throughout this book. Thus far, attention has focused on citizenship strategies of local residents whose actions were situated in the spaces of the neighborhood and were actively circumspect in nature. However, in this moment, for Muslim citizenship to be effectively reclaimed, both for the *Maulānā* in real terms and for the neighborhood in the symbolic sense, *collective* and public action was necessary. Through this "act of citizenship," Madanpura's residents visibly constituted themselves as activist citizens. This public assertion of citizenship contrasts with the strategies explored in Chapter Four, where residents typically pursued more hidden or alternative forms of citizenship. The contrasting orientation of citizenship practices within different times and spaces highlights the ways in which the visibility of Madanpura's residents as a "community" was sometimes deemed desirable or even necessary in order to advance claims to citizenship, whilst at other times being in the public eye was thought to hamper their efforts towards realizing equality and justice, especially where this risked destabilizing everyday peace. Earlier chapters have examined the interacting role of agency and "peace talk" in reproducing both real and imagined geographies of intercommunity engagement. Similarly, it is interesting here how public demands for citizenship were framed in narrative terms of "civility," "solidarity," "kindness" and the "good" citizen, which directly contested notions of the "dangerous terrorist" and served to legitimize claims to citizenship through an appeal to peaceful social qualities. The imperative to protect everyday peace was a central concern in other ways, as Muslim residents were acutely aware that if relationships broke down around the protest, these peaceful demands would be entirely undermined.

Yet, this chapter shows that claims to citizenship and commitments to everyday peaceful sociality have limits within the Indian social and state setting. As this book has shown, everyday peace is a precarious process that is reproduced through the suspension rather than resolution of tensions. And, herein lies the challenge for this act of citizenship to contribute to a type of longer-lasting transformative politics. The protest, in and of itself, contests the notion of the "helpless Muslim victim" whilst seeking to challenge the dominant image of Muslims as "uncivil terrorists" in figurative and physical space. That the Uttar Pradesh police administration responded by securing the *Maulānā's* release could be recognized as a potentially transformative experience in Muslim–police relations, if only temporarily. Indeed the police response did inspire a longer-lasting legacy of satisfaction and confidence amongst Muslim residents, who perceived their claims to have been recognized as legitimate. However, rather ironically, even whilst the protest sought to overturn the notion of Muslim incivility it depended upon the circulation of the idea of the potentially "uncivil Muslim," and the possibility that disorder and disruption could result. Indeed, it was this threatening image of the Muslim "Other" that prompted the police action in the first place and ensured the continuation of everyday peace. Therefore, even whilst the act of citizenship was answered, in many ways the public protest actually worked to *re*constitute Indian Muslims as members of a particular "community" occupying peculiarly Muslim space. In the longer term, the protest represented a material reminder of the perceived threat posed by this Muslim "community" to disrupt everyday urban life. Developing Secor (2004) and Staeheli (2008) this chapter exposes the contradictions of community visibility for India's Muslims, which can simultaneously empower but also undermine the potential for practices of citizenship to be realized.

Finally, this chapter has developed a perspective on the situated geographies of citizenship and the scalar politics of peace. That this protest came to represent an "act of citizenship" in which actors were constituted as citizens and claims were answered, was contingent on a range of factors that operated across different sites and scales. It is very likely that had a different set of events cohered in place in response to the protest, the police might not have responded as sensitively to the protestors' demands; contest may have turned into dissent and the city's peaceful relations may have been disrupted. In such an event, the protestors would have been constituted as criminal subjects and not as citizens. This moment reinforces the importance of vertical relations with state and geopolitical practices and imaginations as well as horizontal relations within society which all inform the potential and imperative for reproducing peace *in situ*. The citizenship strategies articulated here therefore both complement and contrast with those documented in Chapter Four and further demonstrate how local Muslims employed different types of citizenship strategies within different times and spaces, but always in dialogue with the precarious reality of everyday peace.

Note

1   Also known as the monsoon month, Sawan is the Hindu calendar month which falls around July–August. It represents one of the holiest months in the Hindu calendar and, because it is dedicated to the worship of Lord Shiva, large numbers of devotees come to Varanasi on pilgrimage during this time.

Chapter Eight
Conclusions: Questioning Everyday Peace

On September 22, 2010 the neighborhood of Madanpura once again become
the focus of events orientated towards maintaining peace (*shanti rakha hai*) in
Varanasi. In anticipation of the Allahabad High Court's long awaited judgment on the Ayodhya issue, a group of Hindus and Muslims participated
together on a "Peace March" through the streets of Madanpura and its neighboring *mohallās*. The aim was to prevent tension from arising between the two
communities ("*koi tension nahi ho paya*"). In the days immediately before the
announcement, informants observed how a strange (*ajab*) atmosphere had
descended on the city. People were worried about the imminent verdict, and
the implications it might have for Hindu–Muslim relations in the city,
and nation more widely, especially if it appeared to favor one community
over another. These fears were no doubt underpinned by the heavy presence
of police and army personnel on the streets, drafted in especially during this
period. Through their conversations with local residents it became apparent
to friends Farhan and Sunil, Muslim and Hindu respectively, that rumors
were circulating in Hindu neighborhoods about how Muslims had started to
stockpile grains and essential items. The understanding was that Muslims
must be planning "something," whilst Muslims were under the impression

Everyday Peace?: Politics, Citizenship and Muslim Lives in India, First Edition. Philippa Williams.
© 2015 John Wiley & Sons, Ltd. Published 2015 by John Wiley & Sons, Ltd.

that Hindus were planning "something." In an effort to assuage the mistaken feelings of suspicion and mistrust that were beginning to surface, Farhan and Sanjay decided to organize the Peace March as a public statement about their good relations. Both men were already well known in their neighborhoods so when they called on family and friends for support, 144 people turned out to participate and communicate to the local community that, whatever the forthcoming verdict, relations would remain peaceful.

As it happened, the decision was deferred by almost a week due to security concerns at the national level. But, when the judgment was finally delivered on September 30, 2010 for the land under dispute to be divided three ways, with two-thirds going to Hindu parties, the Infant Lord Ram, represented by Hindu Maha Sabha and Nirmohi Akhara and one-third to Muslim actors, the Sunni Waqf Board, peace did persist across the country. Local Madanpura residents were struck by the apparent imbalance of the decision, and the impracticality of enforcing the land divide between unrelated parties. There was widespread agreement in the neighborhood that the decision had not delivered India's Muslims adequate justice, based as it was on articles of the Hindu faith rather than facts (see Roy 2010). Consequently, it was argued that the case should therefore be put to appeal at the National Supreme Court where Muslims understood that justice would be realized. Despite their bla-tant disappointment with the verdict, local Muslims chose not to publicly voice their dissent, a trend that was mirrored elsewhere in north India. Instead, there was a degree of acceptance that the decision, although inherently wrong, was in fact right for the country. As one Muslim friend commented: "consid-ering the situation of the country and the potential for struggles then it is the better judgment... If the verdict was to have come the way it should have, in favor of Muslims, then it would have put India back 20 years due to fights and riots." Amongst local Muslim friends there was a strong feeling that the decision should not be turned into a political cause, anymore than it was already, for the Supreme Court would eventually reach the right decision. The only concern was that this might take another 20 years.

What do these events and thoughts tell us? Talk about "peace" or *shanti* is central to everyday life in Varanasi, it both reflects and shapes relations and encounters between Hindus and Muslims in the city (as do past and poten-tial practices of violence). "Peace talk" and actions towards peace are inevi-tably sharper during periods of tension or threat that play out across different sites and scales, as the urgency for peace to be maintained heightens. As I have argued, everyday realities of peace between Hindus and Muslims deserve excavating, in the same way that processes of violence have more often been subject to in India (e.g. Roy 1994; Nandy 1995; Tambiah 1996). Through the course of this book I have tried to unpack what peace looks like, how it is both socially and spatially produced and reproduced, and situated within a particular cultural political economy. Whilst peace may be portrayed as a utopian condition that is without, or after, politics and violence, the stories imparted here show how peace is political, in and of itself. Understood as an inherently relational construction, peace is both the product of differences and connections, and the context through which they are assembled and negotiated through different techniques of power. With the focus on a Muslim weaving community in Varanasi I have been principally concerned with understanding how this particular subaltern group experience, construct and reproduce everyday peace from within positions of marginality. Importantly, everyday peace is revealed to be constituted through uneven practices of power vis-à-vis the state and society, which inevitably come to bear upon the potential for articulating and realizing citizenship. This concluding chapter reflects on the key contributions of the book for thinking about the micropolitics of everyday peace and the practice of citizenship for a particular section of India's Muslims. In so doing it questions the possibility of "everyday peace" and forces a more nuanced understanding of the uneven geographies of power through which peace is continually produced and reproduced.

The Politics of Everyday Peace

Since the early 2000s, political scientists conducting research in South Asia have turned their attention towards understanding not just why violence happens between Hindus and Muslims, but also why it does not happen (Varshney 2002; Brass 2003). In human geography, scholars have sought to eschew violence as a legitimate means to political ends (Kearns 2009; Megoran 2011), and to displace it as the central focus of enquiry in order to uncover the more prosaic and often hidden practices that take place in the context of violence (Hyndman 2001; Pain 2009). Pertaining to the local, anthropologists have also explored the ongoing dynamics of peaceful sociality (Briggs 1994; Fry 1994; Bonta 1996; Graham and Fry 2004) and the maintenance of peaceful exchanges in potential conflict situations (e.g. Bailey 1996; Ring 2006). Critical international relations theorists, meanwhile, have started to think about the local ramifications of peace and the role of agency, in conjunction with perspectives on national and geopolitical peace projects (e.g. Richmond 2009). This book both complements and builds upon these accounts of peace, but it engages more explicitly and closely with the assertion that peace is political. Whilst peace may be seen to persist and contain the potential for justice and equality, as a process, it more often conceals and perpetuates uneven relations of power. My account has been engaged with questions of everyday urban interactions between Muslims and Hindus. The spaces of everyday city life – the neighborhoods, streets, markets and buildings – provide the arenas through which peaceful relations find expression across difference. But these spaces are simultaneously the outcomes – what is at stake – in efforts towards maintaining peaceful relations. I focus special attention on the spaces of the silk sari market, the Muslim *mohallā* and its structures, and how these become reconfigured in line with shifting social arrangements (through time). The sari market is composed of a network of decentralized urban spaces, constituted predominantly by Muslim laborers and producers who circulate around the main sari markets at Chowk, run largely by Hindu wholesale traders (see Chapter Six). It is therefore an arena of cooperative intercommunity relations, at the same time as its existence critically rests upon the perpetuation of local peaceful dynamics. Underpinned by a sense of shared economic fate these spaces of economic peace were constructed in opposition to the city more generally, where Hindu–Muslim relations were regarded as simultaneously more volatile, and less likely to quickly resume in the case of conflict. Beyond the sari market, Varanasi was widely identified as a city of "communal harmony"; local residents took collective pride in this reputation for comparative peace which was spontaneously evoked when comparing life in Varanasi to elsewhere in India.

Urban spaces of peace are therefore dynamic and open to ongoing contest and disruption in the context of local, national and geopolitical events. In light of which, I have attempted to understand the processes and efforts towards maintaining peace within socially constituted spaces. Chapter Five illustrated how the material and imagined boundaries between the Muslim *mohallā* of Madanpura and its neighboring Hindu areas became progressively reified with the local and national rise of Hindu nationalist forces. In this setting, notions of security and insecurity were differentially constructed by different parties, and sometimes intentionally manipulated to further political ends. The annual *pujā* processions by the Bengali community through Madanpura functioned to disrupt everyday peace and demanded the considerable deployment of police bodies and infrastructure within the neighborhood. These events revealed contrasting interpretations of threat, security and insecurity. The social production of policed space incited notions of security for the Bengali communities, but insecurity for Madanpura's Muslims, for whom the Indian administration had typically represented the face of violence and discrimination, rather than shelter and protection. This scenario emphasizes that there are different sides to peace; efforts to maintain peace are inevitably designed in a particular image such that some groups come to experience freedom and security whilst others suffer incarceration and fear.

The relationship between space and peace was also imaginatively represented and constituted, at once circulated, produced and reproduced through the articulation of narratives. Within the spaces of the nation, the idea of secularism, underpinned by the quest for "unity in diversity," informs the aspirational framework within which relations between religious communities are performed and shaped. The potential to simultaneously recognize and live together across difference underscored the local narratives of peace that came to portray relations within Varanasi's urban publics. Given the nature of economic interdependence within the silk sari industry, Hindu and Muslim informants alike, referred to a style of Hindu–Muslim *bhāī-bhāī* or *bhaīachārā* that characterized life in the city. The idea of intimacy and shared heritage that this evoked was locally grounded within the urban and animated through a common identity, as Banarsis. This was situated and further reinforced within the wider cultural political economy of the purvanchal region, where the notion of Hindu–Muslim conviviality was encapsulated in the term *Gangā Jamuni tahzēb/sanskritī*. It is noteworthy that the perception of Hindu–Muslim *bhaīachārā* proved more widespread than the experiences of intercommunity familiarity it purportedly reflected. However, as imagined geographies of peace these ideas proved to be important rhetorical devices in reproducing lived spaces of everyday peace.

Emanating from within the spaces of the silk sari industry, the idea of Hindu–Muslim brotherhood was heavily invested in, even by those who were more ambivalent about their actual experiences of *bhaīachārā*, whether Hindu or Muslim, whether they worked in the sari industry or not. As I argue most animatedly in Chapters Three and Six, narratives that characterized peaceful relations, or "peace talk," both reflected and maintained the status quo. The frequent talk of everyday Hindu–Muslim brotherhood served to depoliticize experiences of difference and discrimination and eschew the possibility of violence. As a normative discourse, the idea of everyday peace may have deferred tensions between these two communities, but it did so by forestalling questions about everyday Muslim experiences of inequality and injustice vis-à-vis the majority Hindu population. In Varanasi, the persistence of peace was therefore contingent on repudiating rather than excavating differences between Hindus and Muslims, businessmen and weavers.

How "peace talk" not only finds expression but also resonates within particular spaces and moments concerns, in part, the role of actors and the question of social agency. Brass (2003) has rightly drawn attention to the importance of agency in understanding the chain of events that lead to Hindu–Muslim violence. Whilst he would perhaps view human action in the context of "ethnic" differences as naturally orientated towards violent exchange, unless otherwise prevented, the position extended in this book is that different actors do work collectively towards peaceful outcomes; why and how this plays out deserves scrutiny. The disruptions to everyday peace caused by the terrorist attacks (Chapter Three) and the *pujā* processions (Chapter Five) offered distinct settings through which to explore not only which actors were implicated in the project of maintaining peace, but also how their actions were interpreted and responded to, or not.

The Indian national and state governments have, in the past, been blamed for not doing enough to maintain peace in the face of potential conflict, and at other times for actually exacerbating tensions and fuelling Hindu–Muslim violence (Varadarajan 2002; Wilkinson 2004). In the mid to late 2000s, when research for this book was being conducted, the Congress led UPA government was in power at the center, and in Uttar Pradesh the caste-based Samajwadi Party held power from 2003 until 2007, when the Dalit party, the Bahujan Samaj Party won a landslide victory. The fractured terrain of party politics in UP and the weakened condition of the Hindu right party, the BJP, informed a broader context in which both the motivation and potential for inspiring anti-Muslim politics, reminiscent of the mid to late 1990s, was not similarly available. Accordingly, faced with a situation of anticipated conflict and violence following the terrorist attacks, party political actors were primarily concerned with working towards, rather than undermining inter-community peace. Yet, city residents from all backgrounds greeted actions by the state or *sarkar* with suspicion. In order to see peace maintained, the local inclination was to keep the state at a distance and give precedence to the gestures of local agencies, notably the *Mahant* of the temple hit by the bomb attacks, and the *Mufti* who also came to the fore during this period.

There is an emerging trend within critical international relations to center the role of local agencies in negotiating routes to post conflict peace (Richmond 2009, 2010). But, in order to understand why some actors might prove more influential than others at restoring or keeping the peace, I argue that due attention should also be given to the question of legitimacy. Different agencies are variously shaped by and granted legitimacy within particular cultural political economies. The reality of legitimacy is a precarious and contingent one, which makes it difficult to determine. What emerges in the course of the book is that politicians and the police, who often acquire their legitimacy through force and power earned through divisive practices, were either sidelined or treated with mistrust by one or both communities in the context of maintaining peace. The persistence of peace was therefore not contingent on the application of force and top-down interventions by state actors. Rather, the possibility for everyday peace resided in a local capacity to create real and imagined spaces of connection, tolerance and civility.

The actions of both local actors and informal, rather than state determined, peace committees were understood to be authentic, and therefore legitimate, because they were embedded within local networks and spoke to vernacular narratives of peace. The authenticity of power and influence is recognized by both Michelutti (2008) and Shah (2010), who conclude respectively that the vernacularization of caste politics and familiarity with indigenous systems of political power grant particular practices and actors in India the authority to build trust and to act and shape their surroundings and social relations.

The question of agency also arouses concerns about responsibility, and who is accountable for maintaining peace. This matter is usually addressed from a geopolitical and/or national perspective involving unilateral and national actors (Bellamy 2009; Thakur 2011) and concerns thinking about different ethical positions (Nardin 1995; Pangle and Ahrensdorf 1999). As an ordinary, peaceful city the matter of actively maintaining peace was not an overt and daily priority for the state machinery in Varanasi. However, as and when tensions and the potential for violent conflict did apparently arise, the police administration, local governments and civil defense committees were visibly responsive. But, even whilst these state actors took responsibility for maintaining peace, they often lacked the necessary legitimacy and trust among local residents, in order to be effective. In this context, I argue that the responsibility for reproducing peace appeared to rest with ordinary city residents, and more substantially on the shoulders of the city's Muslims, who responded through strategies grounded in pragmatism and resilience, which were underscored by degrees of acceptance for their situation. This situated politics of peace was therefore intimately related to the potential for Muslims to articulate and realize their citizenship practices, and to experience social change and transformative political potential.

Muslim Citizenship, Justice and Transformative Possibilities

By turning attention to the complexities of everyday peace, this book has focused on the ways in which a section of India's Muslim minority came to negotiate everyday urban relationships of difference and discrimination as well as connection and cooperation. In this context I have questioned to what extent and in what ways Muslims articulated citizenship practices. Understood as a bundle of practices that surround relationships between the state and individuals, citizenship is intimately concerned with both the political and the social. I argue that, like peace, citizenship is an ongoing process, which is always in a state of becoming rather than a tangible outcome. Here, I have attempted to show how the practice of negotiating and realizing citizenship for India's Muslims was intimately linked with the art of maintaining peace, within a particular cultural political economy. Urban citizenship often concerns everyday struggles over identity, difference and belonging between real and imagined neighbors, as well as the state. These struggles are articulated in and through multiple scales, and differentially inflected and informed by histories of violence and nonviolence.

As I remarked in the Introduction to the book, it is rare for political institutional ideals concerning citizenship to find their mirror image within society. More often, actual practices of citizenship are disjunctive, contested and only partially realized in everyday life, especially by marginalized groups (Isin 2002; Chatterjee 2004; Holston 2008). The possibility of enacting citizenship is in reality contingent on one's location within the social, as spaces of inclusion and exclusion are produced through social processes and uneven power relations (Isin 2008). In India, the majority Hindu community has dominated the construction of citizenship narratives. Although shifting, these have often functioned to symbolically and materially exclude India's Muslims, through both time and space. I contend that in contemporary India, Indian Muslims are consistently portrayed as "helpless victims" or "dangerous terrorists." These narratives serve as representational shorthand for a community that, as "victims," are universally discriminated against, devoid of social agency and forced to segregate themselves and withdraw from mainstream society. And, as "terrorists," they are constructed as constituting a material and ideological threat to the nation's peace and security. Such narratives function to depoliticize the dominant structures of power by obscuring the persistent structural inequalities that really shape Muslim lives, and thereby simultaneously deny the possibility of the Indian Muslim citizen. Yet, as the chapters in this book have shown, in spite of, or perhaps because of, the impartial realities of citizenship, and everyday practices of material and symbolic exclusion, India's Muslims do find ways of realizing citizenship, positively engaging with the state and society, and articulating constructive social agencies. Furthermore, I argue that practices of citizenship are more often orientated towards reproducing secularist ideals and maintaining everyday peace.

As Chapters Four and Seven documented, Muslim practices of citizenship were strategic, sometimes visible and sometimes invisible, but necessarily informed by the cultural political economy in which they were entangled. Given the state's absence in the *mohallā* with respect to welfare provision, and its failure to guarantee justice on the ground concerning the opening of Dr. Massoud's mosque, Muslim *Ansāris* demonstrated degrees of acceptance and pragmatism. Whilst they blamed the state and recognized its prejudiced behavior towards the minority community, they were mindful that publicly protesting against the state's neglect as a "community" would probably invite unwelcome opposition from Hindu neighbors in the city and, in any case, it was unlikely that the municipal or state government would be capable of responding to their requests. Instead, Muslim *Ansāris* articulated more invisible forms of citizenship by establishing autonomous, alternative spheres for delivering welfare provision, not only to Muslim residents, but also to poorer socioeconomic groups from across the city. Such actions rebutted the construction of the Muslim as "helpless victim." These citizenship practices were framed by narratives of hospitality and shared national identity and animated the way that secularism, despite its drawbacks, nonetheless offered opportunities for Muslims to realize their citizenship and to participate meaningfully in the nation and everyday city life. They also pointed to forms of "cosmopolitan neighbourliness" (Datta 2012), where residents made active attempts to form bridges across religion and counter oft-made claims that Muslims chose to segregate and isolate themselves within India's cities.

By contrast, the arrest of Maulana Abdul Mateen provoked a far more visible act of citizenship by local residents. Chapter Seven shows how they collectively staged a public protest within the neighborhood in order to ask questions about the perceived (in)justice and demand answers from the police administration. In this scenario, assembling, as a "community" within public space, was deemed necessary in order to advance claims to citizenship. Both the *Maulānā's* freedom and the reputation of the neighborhood more generally rested upon the police administration overturning their decision and releasing the *Maulānā* without charge.

By drawing other residents, the police and politicians into the scene of the protest and asking questions about justice, this act of citizenship may be deemed political. But, it was also ethical because it elicited a particular response from those with whom it engaged. As answerable, it was paramount that the protest remained civil and that everyday peace was not undermined in the process. The events underscore the important role of actors and the kind of labor that was necessarily invested in actually realizing claims to citizenship and maintaining ongoing peaceful relations in the city. These contrasting scenarios develop a picture of multiple and plural Muslim citizenship practices. As strategies enacted simultaneously within different sites and spaces, as well as through time (Shani 2010), citizenship practices were sometimes highly visible and engaged directly with questions of justice, whilst at other times they were more hidden and less political.

Given my research focus on predominantly public spaces of intercommunity encounter and exchange my informants proved to be mostly male, and the mechanisms towards peace as well as citizenship practices outlined here tended to be constructed by men and male narratives. The gendered nature of such practices is striking and reflects both my research focus, but also an important grounded perspective on everyday intercommunity relations. The positive influence of men in constructing and reproducing spaces of everyday peace contrasts with debates in the peace studies literature which typically conceive of women as "natural" peacemakers and peacebuilders (Diop 2002; Charlesworth 2008). Whereas men tend to dominate the spaces of formal institutions, female agency is rooted in the community and orientated towards bridging relations. Consequently, it is argued that the potential for female participation in peacebuilding has been underutilized in postconflict scenarios (Anderlini 2007). The particular nature of the cultural political economy in Varanasi – the centrality of economic spaces in facilitating inter-community interaction and Islamic cultures of *purdah*, which limit both fleeting and sustained intercommunity engagement for females – means that male interaction typically structures the material and rhetorical practices of everyday peace and citizenship in this city. How this gendered experience of everyday peace and citizenship impacts upon Muslim women in the city should be the subject for further research.

In the context of reproducing everyday peace and articulating citizenship, "the state" appears omnipresent, both through its absence, as well as its presence. Widely perceived by Madanpura's residents as prejudiced against their community, in terms of its failure to deliver welfare and history of symbolic and material violence against them, the state was largely treated with a degree of skepticism and prudence. Yet, it was also regarded as desirable, and expectations continued to be invested in the state to deliver. As a set of organizations and as an idea, "the state" has attracted considerable attention in India (see Williams et al. 2011). The conflicting experiences of Muslim *Ansāris* in Varanasi reflect the view of the state put forth by Hansen (2001), as simultaneously "profane," meaning dark and incoherent, and "sublime," referring to its capacity to reside above society and represent the epitome of rationality and justice. But, for a community that has experienced persistent and pervasive patterns of discrimination within society, more importantly, the state offered a last resort. However, "the state" was differentially interpreted. It was accepted that the lower levels of the state architecture, such as the police administration and government officials who were more embedded within everyday social relations, would not fully deliver, and sometimes even contest, Muslim rights in practice. Conversely, Muslim expectations for the courts and judiciary to eventually deliver their constitutional commitments to secularism and democracy were unswerving. The opening to this chapter reiterates that sentiment, following the recent judgment made by the Allahabad High Court concerning the apparently disappointing division of land at Ayodhya, local Muslims continued to have ultimate faith in the potential of the national Supreme Court to hear an appeal and finally deliver real justice. Importantly, the persistence of hope and patience, despite widespread experiences of discrimination, underpins Muslim engagements with the higher reaches of the state.

Engaging with the question of Muslim agency in everyday urban settings my findings also show how, outside of formal political channels, mediations between different citizens and the state take place that may promote connections and experiences of citizenship (see Corbridge et al. 2005, p. 257). Unlike the typically illegal and violent spaces of engagement that characterize the notion of "political society" outlined by Chatterjee (2004), I have documented the creative, generally legal, nonviolent ways in which Muslim agencies applied themselves in order to maneuver for their share of social and economic resources. These activities have been centrally concerned with establishing complementary civic publics, reproducing civil spaces and protecting the community's reputation as comprising civil individuals and above all good Indian citizens. In this context, I argue that Muslim residents articulated the capacity as well as the need, as marginalized "citizens," to expand their citizenship opportunities by producing alternative "civic" spaces and also struggling for rights and justice *within* what Chatterjee terms "political society" (see also Jeffrey 2010; Harriss 2010).

Threaded through the different strands of this book is the matter of justice and, more appropriately, *in*justice. By and large, notions of injustice were interpreted in relational terms, such that Muslim weavers typically conceived of their situation in relation to the majority Hindu population, rather than against the ideals of the state. What is interesting is that experiences of injustice rarely provoked actions that were explicitly orientated towards realizing *maximum* justice, perfect justice, or which openly contested or resisted the structures of power. Rather, agencies were more often articulated in an attempt to minimize injustices enacted against them, and almost always in ways which upheld the laws of the land and actively sought to reproduce rather than undermine everyday peace. How notions of injustice/justice were approached informed the kind of citizenship strategy that was pursued by local Muslim residents.

The nexus between citizenship strategies, the reproduction of everyday peace and a pragmatic approach to injustice/justice informed the potential for transformative politics in local Muslim lives. In light of the findings documented throughout this book, it is apparent that whilst Varanasi's Hindus and Muslims came together in both predictable and more periodic ways within different urban publics, the nature and implication of this social contact varied. The silk sari industry may have engendered the kind of interdependent "micropublic" advocated by Ash Amin, however whilst intercommunity relations were structured by sentiments of tolerance and respect on the surface, it was apparent that prejudices and patterns of inequality more often persisted. As the religious minority, Muslims more often accepted daily injustices and their place as those to be tolerated, whilst the majority Hindu community quietly expressed a sense of their generosity for extending feelings of tolerance, and sometimes protection to the minority. There were instances of sustained intercommunity friendships within the silk sari industry and the *mohallā*, yet these more transformative encounters were experienced at the individual level, and did not influence wider group experiences (Valentine 2008).

More generally, for India's Muslims, the potential for transformative politics was significantly constrained within a national political landscape that has framed matters of social justice, particularly concerning access to socioeconomic and educational rights, in caste terms (see Yadav 2009). At the same time, the promotion of rights on the basis of religious identity represents an acutely contentious area for government and public debate. The prominence of the Hindu right within political circles continues to inspire a rhetoric of minority appeasement which works to limit the political expediency of actually addressing patterns of inequality between India's Muslims and other socio-religious groups. Public and political interest around the condition of India's Muslims (e.g. Prime Minister's High Level Committee 2006) may constitute an important juncture in thinking through the condition of India's largest religious minority. However, the agency of India's Muslims and those working on behalf of them continues to be circumscribed in particular ways.

Nongovernmental organizations (such as the People's Vigilance Committee on Human Rights), concerned with improving the condition of weavers in the Varanasi region, the majority of whom are Muslim, were forthright about the options available to them. Cognizant of communally rooted tensions within the silk sari industry and the city more widely, the organization's director was adamant that campaigns towards the uplift of weavers should be framed in secular terms that drew attention to occupational and economic circumstances, and played down the majority weavers' religious identity. Whilst this secular approach serves as an apparently strategic campaigning tool, it also highlights the irony underpinning the wider reproduction of everyday economic peace and of Muslim marginalization that is rooted in a particularly resistant form of Hindu hegemony. Harriss-White (2005) was right to be anxious about the economy's vulnerability to religious influences, especially where practices of power, in the case of Varanasi's silk sari industry at least, are perpetuated under the veneer of secularism and everyday peace, locally expressed as Hindu–Muslim *bhaīachārā*/brotherhood.

Geography and Everyday Peace

An important value of this book rests on its detailed attention to everyday peace in a particular urban and regional setting, however it also engenders wider theoretical significance for interpreting peace as something "real" in its own right. It responds to calls from the field of peace studies to develop empirical studies of peace from the "ground up," as it actually happens, rather than extend theories of how peace should be (see Jutila et al. 2008; Richmond 2008; Gleditsch 2014). The book has mobilized the concept of "everyday peace" as a framework through which to witness and interpret the prosaic spaces of peace in a relatively "ordinary" peaceful city. One of the book's key questions concerned how peace is socially and spatially reproduced. The notion of everyday peace serves important analytical purchase by turning the lens on to peace in a locality and the experiences of different people as they negotiate their everyday lives. This approach counters romantic conceptualizations of peace that are disembodied and detached from the geographies of place. Just like violence, peace sits in sites, spaces and bodies too, which are intertwined with practices and discourses across scale. Conjoining peace and the everyday therefore opens up the potential to understand how peace makes place and place makes peace. This view contrasts with popular interpretations of the city as the primary "battlespace" (Graham 2006) and goes some way to showing how cities can also "'make peace' as a condition of enduring harmony, rather than merely create an absence of war" (Fregonese 2012, p. 297).

Situating peace in the everyday and in place draws attention to the role of agency, by different actors such as religious leaders, civil society actors, politicians and the police as well as ordinary city residents. This has highlighted questions concerning contextual contingencies, legitimacy and responsibility which are vitally important to understand. Not only why some agencies chose to act constructively in some times and places and destructively in others, but also why some actors are more effective at keeping the peace than others. And, it prompts us to witness agencies for peace not just in the conscientious and conspicuous "peacekeepers" but also through ordinary actors whose everyday role in keeping the peace is less consciously and visibly expressed.

In asking the question, "What does peace look like?" this book has offered a view of peace as process, as a condition that is always becoming, under negotiation and demanding of ongoing labor. Understanding peace as process is not novel *per se*, yet peace as an everyday process is rarely understood in less dramatic environments away from recognizable postconflict zones. Even in relatively peaceful places peace is not void of tensions and conflict. Indeed, the memories and the potential of material violence significantly inflect the everyday reality of peace. What is shown here is that articulating everyday peace from the margins through expressions of acceptance, resilience and pragmatism in themselves represent forms of resistance to violence, and the potential injustices it perpetuates and creates.

In developing geographical concerns with peace, the book has shown how everyday peace is reproduced through both material *and* imagined geographies of everyday encounter. Importantly, practices of encounter and coexistence are unevenly experienced, as interactions between different communities shift through space and time. It is striking that everyday peace is re-made through encounters in the city which can be both constructive, embodying trust and friendships, and in other spaces destructive, involving practices of intimidation and coercion. This illustrates the uneven, sometimes uncertain quality of everyday peace which informs the precarious reality of everyday peace. What the book has also shown is that experiences of everyday intimate and positive encounter do not have to be universal for everyday peace to be popularly imagined. "Peace talk" may reflect a degree of reality, but more importantly it contains an expectation and belief in the possibility to re-make peace in a particular place.

Examining peace through a critical lens has exposed the politics of peace, and powerfully challenged theoretical conceptions of peace as a space after or without politics. This reinforces the imperative for more studies to examine peace in its own right from the "bottom up." Recognizing the politics of peace informs and empowers attempts to discover in whose image peace is reproduced and how different people experience peace differently. As Richmond (2008) has argued, the dangerous potential of peace is that it is valued as an unprecedented good, without examining the processes through which peace is continued. Understanding the politics of peace from the margins highlights the uneven geographies of power that inform encounters and the reproduction of peace.

I argue that citizenship provides a framework for interpreting the scalar politics of peace, which takes into account the interaction between horizontal relations with "society" and vertical relations with "the state." It shows how ongoing strategies for citizenship and everyday peace from the margins are carefully negotiated and often rooted in similar struggles for inclusion. And, where being "included" typically entails not challenging the status quo then it follows that reproducing everyday peace in the real world, is not contingent on realizing perfect justice. To the contrary, aspirations for ultimate justice are often conceded or superseded in order to prevent tensions and conflict and safeguard everyday peace. As Amartya Sen (2009) argues, it is important to interpret justice as it is actually lived and negotiated, along a continuum, from possibly less unjust to more just realities, rather than against the idea of a perfectly just society. Conceptualizing everyday peace as less-than-just sits uncomfortably with Galtung's notion of "positive peace" (1969), which requires the presence of absolute social justice. As it actually happens, situated peace is more untidy, unequal and unjust, but it is also constitutive of positive encounters and relations that are not entirely captured by the "absence of violence."

Everyday peace offers an analytical framing for understanding how peace, as a sociospatial relation, is reproduced through and against different sites and scales. As Koopman persuasively argues, we need to think of peace as "located and spatial, as practical and material and, as such, as necessarily plural" (2014, p. 111). In *this* situated account of peace uneven experiences of citizenship, justice and inclusion underpin the shifting fragility of peace and are integral to its re-making in place. Yet, the value of understanding situated knowledges of different peace lies in building a picture of the contrasts, continuities and connections between different peaces, in different places. How is "peace talk" constituted in other socioeconomic and historical contexts? Who "peoples" peace? How are matters of responsibility and legitimacy constructed and realized? How are practices and agencies of peace located within changing (geo)political realities? What is the relationship between citizenship, justice and peace in other places, for other people? Proliferation of these kinds of critical and diverse conversations about the ongoing realities of peace will play an important role in how we construct the value and potential of our societies, now and in the future.

References

Abdullah, H. (2002) Minorities, education and language: the case of Urdu. *Economic and Political Weekly* 37(24), 2288–2292.

Abu-Lughod, L. (1986) *Veiled Sentiments: Honor and Poetry in a Bedouin Society.* University of California Press, Berkeley.

Abu-Lughod, L. (2008 [1990]) The romance of resistance: tracing transformations of power through Bedouin women. *American Ethnologist* 17(1), 41–55.

Agrawal, Y. (2004) *Silk Brocades.* Lustre Press, New Delhi.

Ahmad, I. (2009) The secular state and the geography of radicalism. *Economic and Political Weekly* 44(23), 33–39.

Alam, A. (2003) Democratisation of Indian Muslims: some reflections. *Economic and Political Weekly* 38(46), 4881–4885.

Alam, M.S. and Raju S. (2007) Contextualising inter-, intra-religious and gendered literacy and educational disparities in rural Bihar. *Economic and Political Weekly* 42(18), 1613–1622.

Alatout, S. (2009) Walls as technologies of government: the double construction of geographies of peace and conflict in Israeli politics, 2002–present. *Annals of the Association of American Geographers* 99(5), 956–968.

Amin, A. (2002) Ethnicity and the multicultural city: living with diversity. *Environment and Planning* A 34(6), 959–980.

Everyday Peace?: Politics, Citizenship and Muslim Lives in India, First Edition. Philippa Williams.
© 2015 John Wiley and Sons, Ltd. Published 2015 by John Wiley & Sons, Ltd.

Amin, A. (2012) *Land of Strangers*. Polity Press, Cambridge.

Anderlini, S.N. (2007) *Women Building Peace: What They Do, Why it Matters*. Lynne Rienner, Boulder, CO.

Anderson, K. (1991) *Vancouver's Chinatown: Racial discourse in Canada, 1875–1980*. McGill-Queen's University Press, Montreal.

Ansari, I.A. (2006) *Political Representation of Muslims in India (1952–2004)*. Manak Publications, New Delhi.

Appadurai, A. (1996) *Modernity at Large: Cultural Dimensions of Globalisation*. University of Minnesota Press, Minneapolis.

Appadurai, A. (2002) Deep democracy: urban governmentality and the horizon of politics. *Public Culture* 14(1), 21–47.

Appadurai, A. (2006) *Fear of Small Numbers: An Essay on the Geography of Anger*. Duke University Press, Durham, NC.

Ara, A. (2004) Madrasas and the making of Muslim identity in India. *Economic and Political Weekly* 39(1), 34–38.

Arendt, H. (1969) *On Violence*. Harcourt Books, Orlando, FL.

Asad, T. (1990) Multiculturalism and British identity in the wake of the Rushdie affair. *Politics and Society* 18(4), 455–480.

Assayag, J. (2004) *At the Confluence of Two Rivers: Muslims and Hindus in South India*. Manohar, New Delhi.

Bailey, F.G. (1996) *The Civility of Indifference: On Domesticating Ethnicity*. Oxford University Press, Delhi.

Bajpai, R. (2002) Minority rights in the Indian Constituent Assembly Debates, 1946–1949. *QEH Working paper series* 20, 1–39.

Bakhtin, M. (1993) *Toward a Philosophy of the Act*. University of Texas Press, Austin.

Barash, D.P. (2000) *Approaches to Peace: A Reader in Peace Studies*. Oxford University Press, Oxford.

Barnett, C. (2011) Geography and ethics: justice unbound. *Progress in Human Geography*, 35(2), 246–255.

Barnett, C. and Low, M. (2004) *Spaces of Democracy: Geographical Perspectives on Citizenship, Participation and Representation*. Sage Publications, Thousand Oaks, CA.

Bayat, A. (2008) Cairo cosmopolitan: living together through communal divide, almost. In: Mayaram, S. (ed.) *The Other Global City*. Routledge, London, pp. 179–201.

Bayly, C.A. (1983) *Rulers, Townsmen and Bazaars: North Indian Society in the Age of British Expansion, 1770–1870*. Cambridge University Press, Cambridge.

Beckett, K. and Herbert S. (2010) *Banished: The New Social Control in Urban America*. Oxford University Press, Oxford.

Bellamy, A. (2009) *The Responsibility to Protect: The Global Effort to End Mass Atrocities*. Polity, Cambridge.

Benhabib, S., Shapiro I. and Petranovic D. (eds.) (2007) *Identities, Affiliations, and Allegiances*. Cambridge University Press, Cambridge.

Berreman, G.D. (1963) *Hindus of the Himalayas: Ethnography and Change*. Oxford University Press, New Delhi.

Bhan, G. (2009) This is no longer the city I once knew: evictions, the urban poor and the right to the city in millennial Delhi. *Environment & Urbanization* 21(1), 127–142.

Bhargava, R. (1998) *Secularism and its Critics*. Oxford University Press, Delhi.

Bishop, R. and Roy T. (2009) Mumbai: City-as-Target: Introduction. *Theory, Culture & Society*. 26(7–8), 263–277.

Bismillah, A. (1996) *The Song of the Loom (Jhini Jhini Bini Chadariya)*. Macmillan, Madras.

Blumen, O. and Halevi S. (2009) Staging peace through a gendered demonstration: women in black in Haifa, Israel. *Annals of the Association of American Geographers* 99(5), 977–985.

Bonta, B.D. (1996) Conflict resolution among peaceful societies: the culture of peacefulness. *Journal of Peace Research* 33(4), 403–420.

Borchgrevink, A. (2003) Silencing language: of anthropologists and interpreters. *Ethnography* 4(1), 95–121.

Boulding, K.E. (1977) Twelve Friendly Quarrels with Johan Galtung, *Journal of Peace Research* 14(1), 75–86.

Boulding, K.E. (1988) Moving from unstable to stable peace. In A. Gromyko and M. Hellman. (eds.) *Breakthrough: Emerging New Thinking: Soviet and Western*. Available at: http://www-ee.stanford.edu/~hellman/Breakthrough/book/contents. html, Accessed April 19, 2015.

Brass, P. (1997) *Theft of an Idol: Text and Context in the Representation of Collective Violence*. Princeton University Press, Princeton, NJ.

Brass, P. (2003) *The Production of Hindu–Muslim Violence in Contemporary India*. Oxford University Press, New Delhi.

Brass, P. (2004) Elite interests, popular passions, and social power in the language politics of India. *Ethnic and Racial Studies* 27(3), 333–375.

Brass, P. (2006) Collective violence, human rights and the politics of curfew. *Journal of Human Rights* 5(3), 323–340.

Breman, J. (2002) Communal upheaval as resurgence of social Darwinism. *Economic and Political Weekly* 37(16), 1485–1488.

Brickell, C. (2000) Heroes and Invaders: gay and lesbian pride parades and the public/private distinction in New Zealand media accounts. *Gender, Place and Culture* 7(2), 163–178.

Bringa, T. (1995) *Being Muslim the Bosnian Way: Identity and Community in a Central Bosnian Village*. Princeton University Press, Princeton, NJ.

Brodie, J. (2008) The social in social citizenship. In: Isin E. (ed.) *Recasting the Social in Citizenship*. University of Toronto Press, Toronto, pp. 20–43.

Brubaker, R., and D. Laitin. (1998) Ethnic and nationalist violence. *Annual Review of Sociology* 24(1), 423–52.

Burgess, R.G. (1984) *In the Field: An Introduction to Field Research*. Allen and Unwin, London.

Carroll, B.A. (1972) Peace research – Cult of power. *Journal of Conflict Resolution* 16(4), 585–616.

Casolari, M. (2002) Role of Benares in constructing political Hindu identity. *Economic and Political Weekly* 37(15), 1413–1420.

Chandhoke, N. (1995) *State and Civil Society: Explorations in Political Theory*. Sage Publications, New Delhi.

Chandhoke, N. (2005) Exploring the mythology of the public sphere. In: Bhargava, R. and Reifeld, H. (eds.) *Civil Society, Public Space and Citizenship*. Konrad Adenauer Stiftung, New Delhi.

Chandhoke, N. (2009) Civil society in conflict cities. *Economic and Political Weekly* 44(31), 99–108.

Chandra, M. (1985) *Kashi Ka Itihas*. Vishwavidyalaya Prakashan, Varanasi.

Chandravarkar, R. (1994) *The Origins of Industrial Capitalism: Business Strategies and the Working Classes in Bombay, 1900–1940*. Cambridge University Press, Cambridge.

Chari, S. (2004) Provincializing capital: the work of an agrarian past in south Indian Industry. *Comparative Studies in Society and History* 46(4), 760–785.

Chari, S. and Gidwani V. (2004) Guest Editorial. Geographies of work. *Environment and Planning D: Society and Space* 22(4), 475–484.

Charlesworth, H. (2008) Are women peaceful? Reflections on the role of women in peace-building. *Feminist Legal Studies* 16(3), 347–361.

Chatterjee, P. (1997) Secularism and toleration. In: Chatterjee P. (ed.) *A Possible India: Essays in Political Criticism*. Oxford University Press, New Delhi, pp. 228–262.

Chatterjee, P. (2001) On civil and political societies in post colonial democracies. In: Kaviraj, S. and Khilnani, S. (eds.) *Civil Society: History and Possibilities*. Cambridge University Press, Cambridge.

Chatterjee, P. (2004) *The Politics of the Governed: Reflections on Popular Politics in Most of the World*. Columbia University Press, New York.

Chatterjee, P. (2005) Sovereign violence and the domain of the political. In: Hansen, T.B. and F. Stepputat (eds.) *Sovereign Bodies: Citizens, Migrants, and States in the Postcolonial World*. Princeton University Press, Princeton, NJ.

Chua, A. (2003) *World on Fire: How Exporting Free Market Democracy Breeds Ethnic Hatred and Global Instability*. Doubleday, New York.

Ciotti, M. (2008) "Islam": what is in a name? In: Banerjee, M. (ed.) *Muslim Portraits: Everyday Lives in India*. Yoda Press, New Delhi, pp. 1–10.

Ciotti, M. (2010) *Retro-Modern India: Forging the Low-caste Self*. Routledge, London.

Coaffee, J. and Murakami, D.W. (2008) Terrorism and surveillance. In: Hall, T., P. Hubbard and J.R. Short (eds.) *The Sage Companion to the City*. Sage, London, pp. 352–372.

Cohen, S.E. (2006) Israel's West Bank Barrier: an impediment to peace? *The Geographical Review* 96(4), 682–695.

Connolly, W.E. (1979) Appearance and reality in politics. *Political Theory* 7(4), 445–468.

Corbridge, S. and Harriss, J. (2000) *Reinventing India: Liberalization, Hindu Nationalism and Popular Democracy*. Polity Press, Cambridge.

Corbridge, S., Williams, G., Srivastava, M. et al. (eds.) (2005) *Seeing the State: Governance and Governmentality in India*. Cambridge University Press, Cambridge.

Cortright, D. (2008) *Peace: A History of Movements and Ideas*. Cambridge University Press, Cambridge.

Cowen, D. (2008) *Military Workfare: The Soldier and Social Citizenship in Canada*. University of Toronto Press, Toronto.

Cowen, D. and Gilbert, E. (2008) *War, Citizenship, Territory.* Routledge, London.

Cresswell, T. (1996) *In Place/Out of Place: Geography, Ideology, and Transgression.* University of Minnesota Press, Minneapolis.

Dahlman, C.T. (2004) Geographies of genocide and ethnic cleansing: the lessons of Bosnia-Herzegovina. In: Flint, C. (ed.) *The Geography of War and Peace: From Death Camps to Diplomats.* Oxford University Press, Oxford, pp. 174–197.

Daley, P.O. (2008) *Gender and Genocide in Burundi: The Search for Spaces of Peace in the Great Lakes Region of Africa.* James Currey, Oxford.

Daley, P.O. (2014) Unearthing the local: hegemony and peace discourses in Central Africa. In: McConnell, F., N. Megoran and P. Williams (eds.) *Geographies of Peace.* I.B. Tauris, London, pp. 66–88.

Dalmia, V. (1997) *The Nationalism of Hindu Traditions: Bharatendu Harischandra and Nineteenth-Century Banaras.* Oxford University Press, Delhi.

Darling, J. (2014) Welcome to Sheffield: the less-than-violent geographies of urban asylum. In: McConnell, F., N. Megoran and P. Williams (eds.) *Geographies of Peace.* I.B. Tauris, London, pp. 229–249.

Das, V. (2007) *Life and Words: Violence and the Descent into the Ordinary.* University of California Press, Berkeley.

Das, V. and Poole D. (eds.) (2004) *Anthropology in the Margins of the State.* James Currey, Oxford.

Datta, A. (2012) "Mongrel City": Cosmopolitan neighbourliness in a Delhi squatter settlement. Antipode 44(3), 745–763.

de Certeau, M. (1984) *The Practice of Everyday Life.* University of California Press, Berkeley.

de Neve, G. (2006) Space, place and globalisation: Revisiting the urban neighbourhood in India. In: de Neve, G. and Donner, H. (eds.) *The Meaning of the Local: Politics of Place in Urban India.* Routledge, London, pp. 1–20.

Desai, R. (2002) *Slouching Desai towards Ayodhya.* Three Essays Collective, New Delhi.

Desforges, L., Jones, R., and Woods, M. (2005) New geographies of citizenship. *Citizenship Studies* 9(5), 439–451.

Dikeç, M. (2005) Space, politics and the political. *Environment and Planning D: Society and Space* 23(2), 171–188.

Dikeç, M. (2009) Space, politics and (in)justice. *Justice Spatiale/Spatial Justice.* 1 (December).

Diop, B. (2002) Engendering the peace process in Africa: women at the negotiating table. *Refugee Survey Quarterly* 21, 142–154.

Dodds, K. and Ingram, A. (eds.) (2009) *Spaces of Security and Insecurity: Geographies of the War on Terror.* Ashgate, Aldershot.

Domosh, M. (1998) Those "georgeous incongruities": polite politics and public space on the streets of the nineteenth-century New York City. *Annals of the Association of American Geographers* 88(2), 209–226.

Donald, J. (1999) *Imagining the Modern City.* Athlone Press, London.

Donegan, B. (2011) Spaces for negotiation and mass action within the National Rural Health Mission: "Community Monitoring Plus" and People's Organizations in Tribal Areas of Maharashtra, India. *Pacific Affairs* 84(1), 47–65.

Doron, A. (2007) The needle and the sword: boatmen, priests and the ritual economy of Varanasi. *Journal of South Asian Studies* 29(3), 345–367.

Doron, A. (2008) *Caste, Occupation and Politics on the Ganges: Passages of Resistance.* Ashgate, Aldershot.

Dowler, L. (2001) Fieldwork in the trenches: participant observation in a conflict area. In: M. Limb and C. Dwyer (eds.) *Qualitative Methodologies for Geographers: Issues and Debates.* Arnold, London, pp. 153–164.

Dowler, L. (2002) Women on the frontlines: rethinking war narratives post 9/11. *GeoJournal* 58, 159–165.

Dowler, L. and Sharp, J. (2001) A feminist geopolitics? *Space and Polity* 5(3), 165–176.

Duffield, M. (2002) *Global governance and the New Wars: The Merging of Development and Security.* Zed Books, London.

Dunn, K.M. (2005) Repetitive and troubling discourses of nationalism in the local politics of mosque development in Sydney, Australia. *Environment and Planning D: Society and Space* 23(1), 29–50.

Dwyer, C. (1999a) Veiled Meanings: young British Muslim women and the negotiation of differences. *Gender, Place and Culture* 6(1), 5–26.

Dwyer, C. (1999b) Contradictions of community: questions of identity for British Muslim women. *Environment and Planning D: Society and Space* 31, 53–69.

Dyson, J. (2010) Friendship in practice: girls' work in the Indian Himalayas. *American Ethnologist* 37(3), 482–498.

Eade, J. (1997) Reconstructing Places. In: Eade, J. (ed) *Living the Global City.* Routledge, London, pp. 127–145.

Eck, D. (1983) *Banaras: A city of Light.* Penguin, New Delhi.

Economist. (2008) Ruled by Lakshmi: though inequalities are widening, India's best prescription remains continued rapid growth. *The Economist.* December 11, 2008.

Economist. (2009) Looming extinction. *The Economist.* January 8, 2009.

Eley, G. (1994) Nations, publics and political cultures: placing Habermas in the nineteenth century. In: Calhoun, C. (ed.) *Habermas and the Public Sphere.* MIT Press, Cambridge, MA, pp. 289–339.

Elliot, C.M. (2003) *Civil Society and Democracy: A Reader.* Oxford University Press, New Delhi.

Engineer, A.A. (1995) *Lifting the Veil: Communal Violence and Communal Harmony in Contemporary India.* Sangam Books, Bombay, Hyderabad.

Engineer, A.A. (2003) *The Gujarat Carnage.* Orient Longman, New Delhi.

Engineer, A.A. (2004) Minorities and elections: what are the options? *Economic and Political Weekly* 39(13), 1378–1379.

Enloe, C. (2010) *Nimo's War, Emma's War: Making Feminist Sense of the Iraq War.* University of California Press, Berkeley.

Estlund, C. (2003) *Working Together: How Workplace Bonds Strengthen a Diverse Democracy.* Oxford University Press, Oxford.

Feldman, A. (1991) *Formations of Violence. The Narrative of the Body and Political Terror in Northern Ireland.* The University of Chicago Press, Chicago. Kindle Version.

Felski, R. (2000) The invention of everyday life. *New Formations* 39, 13–32.

Fernandes, L. (2000) Restructuring the new middle class in liberalizing India. *Comparative Studies of South Asia, Africa and the Middle East* 20(1&2), 88–104.

Fernandes, L. (2006) *India's New Middle Class: Democratic Politics in an Era of Economic Reform*. University of Minnesota Press, Minneapolis.

Fincher, R. and Iveson, K. (2011) Justice and injustice in the city. *Geographical Research* 50(3), 231–241.

Flint, C. (2005a) Introduction: geography of war and peace. In: Flint, C. (ed.) *The Geography of War on Peace: From Death Camps to Diplomats*. Oxford University Press, Oxford, pp. 3–18.

Flint, C. (2005b) *The Geography of War and Peace: From Death Camps to Diplomats*. Oxford University Press, Oxford.

Forcey, L.R. (ed.) (1990) *Peace: Meanings, Politics, Strategies*. Praeger, New York.

Foucault, M. (1975) *Discipline and Punish: The Birth of the Prison*. Random House, New York.

Fraser, N. (1990) Rethinking the public sphere: a contribution to the critique of actually existing democracy. *Social Text* 25(25), 56–80.

Fregonese, S. (2012) Urban Geopolitics 8 Years on. Hybrid Sovereignties, the Everyday, and Geographies of Peace. *Geography Compass* 6(5), 290–303.

Freitag, S.B. (1989) *Collective action and Community: Public Arenas and the Emergence of Communalism in North India*. University of California Press, Berkeley.

Freitag, S.B. (1992) *Culture and Power in Banaras: Community, Performance and Environment 1800–1980*. Oxford University Press, Delhi.

Froerer, P. (2007) *Religious Division and Social Conflict: The Emergence of Hindu Nationalism in Rural India*. Social Science Press, New Delhi.

Froystad, K. (2005) *Blended Boundaries: Caste, Class, and Shifting Faces of "Hinduness" in a North Indian City*. Oxford University Press, Oxford.

Fuller, C. and Harriss, J. (2001) For an anthropology of the modern Indian state. In: Benei, V., and C.J. Fuller (eds.) *The Everyday State and Society in Modern India*. C. Hurst, London, pp. 1–30.

Galtung, J. (1996) *Peace by Peaceful Means: Peace and Conflict, Development and Civilization*. Sage Publications, London.

Ganguly, S. and Hagerty, D.T. (2005) *Fearful Symmetry: India–Pakistan Crises in the Shadow of Nuclear Weapons*. Oxford University Press, New Delhi.

Gardiner, M. (2000) *Critiques of Everyday Life*. Routledge, London.

Gayer, L. and Jaffrelot, C. (eds.) (2012) *Muslims in Indian Cities: Trajectories of Marginalisation*. C. Hurst, London.

Ghertner, D.A. (2011) Nuisance talk and the propriety of property: middle-class discourses of a slum-free Delhi. *Antipode* 44(4), 161–87.

Gleditsch, J. (1969) Violence, peace, and peace research. *Journal of Peace Research* 6(3), 167–191.

Gleditsch, Nils Petter, Jonas Nordkvelle and Håvard Strand (2014) Peace research – Just the study of war? *Journal of Peace Research* 51(2), 145–158.

Gooptu, N. (2001) *The Politics of the Urban Poor in Early Twentieth Century India*. Cambridge University Press, Cambridge.

Gottschalk, P. (2000) *Beyond Hindu and Muslim: Multiple Identity in Narratives from Village India*. Oxford University Press, Oxford.

Government of India Census (2001) Uttar Pradesh District Profile. Controller of Publications, New Delhi.

Graham, B. and Nash, C. (2006) A shared future: territoriality, pluralism and public policy in Northern Ireland. *Political Geography* 25, 253–278.

Graham, S. (2006) Cities and the "War on Terror". *International Journal of Urban and Regional Research* 30(2), 255–276.

Graham, S. (2010) *Cities under Siege: The New Military Urbanism*. Verso, London.

Gregory, D (2006) *Colonial Present: Afghanistan, Palestine and Iraq*. Blackwell Publishing, Oxford.

Gregory, D. (2010) War and peace. *Transactions of the Institute of British Geographers* 35(2), 154–186.

Grundy-Warr, C.E.R. (1994) Towards a political geography of United Nations peace-keeping: some considerations. *GeoJournal* 34(2), 177–190.

Gudavarthy, Ajay. (2012) *Re-framing Democracy and Agency in India: Interrogating Political Society*. Anthem Press, London.

Guha, R. (2004) The spread of the salwar. *The Hindu*. October 24.

Gupta, A. (1995) Blurred boundaries: the discourse of corruption, the culture of politics and the imagined state. *American Ethnologist* 22(2), 375–402.

Gupta, D. (1997) Civil society in the Indian context. *Contemporary Sociology* 26(3), 305–307.

Gupta, S. (2006) The Benarasi Weave: separated by religion but united by the spirit of Kashi this duo holds the peace in the holy town. *Outlook*. March 27.

Gupta, S. (2007) The Muslims of Uttar Pradesh. *Economic and Political Weekly* 42(23), 2142–2146.

Hanf, T. (1993) *Coexistence in Wartime Lebanon: Decline of a State and Rise of a Nation*. Centre for Lebanese Studies in association with I.B. Tauris, London.

Hansen, T.B. (1996) Recuperating Masculinity: Hindu nationalism, violence and the exorcism of the Muslim "Other." *Critique of Anthropology* 16(2), 137–172.

Hansen, T.B. (1999) *The Saffron wave: Democracy and Hindu Nationalism in Modern India*. Princeton University Press, Princeton, NJ.

Hansen, T.B. (2001) *Wages of Violence: Naming and Identity in Postcolonial Bombay*. Princeton University Press, Princeton, NJ.

Hansen, T.B. (2007) The India that does not shine. *ISIM Review* 19, 50–51.

Hansen, T.B. (2008) The political theology of violence in contemporary India. *South Asian Multidisciplinary Academic Journal*. 2 'Outrage Communities'. Available at: http://samaj.revues.org/1872, accessed April 19, 2015.

Hansen, T.B. and Stepputat, F. (eds.) (2001) *States of Imagination: Ethnographic Explorations of the Postcolonial State*. Duke University Press, Durham, NC.

Harriss, J. (2005) Widening the radius of trust: ethnographic explorations of trust and Indian business. In: Harriss, J. (ed.) *Power Matters: Essays on Institutions, Politics, and Society in India*. Oxford Collected Essays, Oxford, pp. 167–190.

Harriss-White, B. (2003) *India Working: Essays on Society and Economy*. Cambridge University Press, Cambridge.

Harriss-White, B. (2005) *India's Religions and the Economy*. Three Essays Collective, New Delhi.

Hasan, M. (1988) Indian Muslims since independence: In search of integration and identity. *Third World Quarterly* 10(2), 818–842.

Hasan, M. (1996) The changing position of the Muslims and the political future of secularism in India. In: Sathyamuthy T.V. (ed.) *Region, Religion, Caste, Gender and Culture in Contemporary India*. Oxford University Press, Delhi.

Hasan, M. (1998) *Islam Communities and the Nation: Muslim Identities in South Asia and Beyond*. Manohar, New Delhi.

Hasan, M. (2001) *Legacy of a Divided Nation: India's Muslims since Independence*. Westview Press, Boulder, CO.

Hasan, M. (2004a) *Will Secular India Survive?* ImprintOne, Gurgaon.

Hasan, Z. (2004b) Social inequalities, secularism and minorities in India's democracy. In: M. Hasan (ed.) *Will Secular Indian Survive?* ImprintOne, Gurgaon, pp. 239–262.

Hasan, Z. and Menon, R. (2004) *Unequal Citizens: A Study of Muslim Women in India*. Oxford University Press, Oxford.

Haynes, D.E. (1991) *Rhetoric and Ritual in a Colonial Setting: The Making of a Public Culture in Surat City, 1852–1928*. University of California Press, Berkeley.

Heathershaw, J. (2008) Peace building as practice: discourses from post-conflict Tajikistan. *International Peacekeeping* 14(2), 219–236.

Heitmeyer, C. (2009) "There is peace here": managing communal relations in a town in central Gujarat. *Journal of South Asian Development* 4(1), 103–120.

Held, D. (1989) *Political Theory and the Modern State: Essays on State, Power and Democracy*. Polity Press, Cambridge.

Hertel, B.R. and Humes, C.A. (1993) *Living Banaras: Hindu Religion in Cultural Context*. State University of New York Press, Albany.

Higate, P. and Henry, M. (2009) *Insecure Spaces: Peacekeeping, Power and Performance in Haiti, Kosovo and Liberia*. Zed Books, London.

Hildreth, H. (1981) A house divided: a study of Hindu–Muslim riot that occurred in Banaras in 1977. In: University of Wisconsin. Varanasi: College Year Abroad.

Holmes, R. and B.L. Gan (2005) *Nonviolence in Theory and Practice*. 2nd edition. Waveland Press, Long Grove, IL.

Holston, J. (ed.) (1999) *Cities and Citizenship*. Duke University Press, Durham, NC.

Holston, J. (2008) *Insurgent Citizenship: Disjunctions of Democracy and Modernity in Brazil*. Princeton University Press, Princeton, NJ.

Hopkins, P. (2004) Young Muslim men in Scotland: inclusions and exclusions. *Children's Geographies* 2(2), 252–272.

Hopkins, P. (2007) Global events, national politics, local lives: young Muslim men in Scotland. *Environment and Planning A* 39(5), 1119–1133.

Howell, S. and Willis, R.G. (1989) *Societies at Peace: Anthropological Perspectives*. Routledge, London.

Hsu, Y. (2008) Acts of Chinese citizenship: The tank man and democracy-to-come. In Isin, E. F. and G.M Nielsen (eds.) *Acts of Citizenship*. Zed Books. London

Huberman, J. (2010) Tourism in India: the moral economy of gender in Banaras. In: Clark-Deces, I. (ed.) *A Companion to the Anthropology of India*. Wiley-Blackwell, Oxford, pp. 169–185.

Humphries, C., Marsden M. and Skvirskaja, V. (2008) Cosmopolitanism and the city: Interaction and coexistence in Bukhara. In: Mayaram, S. (ed.) *The Other Global City*. Routledge, London, pp. 202–232.

Hunt, K. (2006) Bismillah Khan: virtuoso musician who introduced the shehnai to a global audience. *The Independent*. August 22.

Hussain, D. (2008) Hindu–Muslim *bhai bhai* in a small town in Bangladesh. *Economic and Political Weekly* 44(20), 21–24.

Hussain, D. (2013) *Boundaries Undermined: The Ruins of Progress on the Bangladesh/India Border*. C. Hurst, London.

Hyndman, J. (2001) Towards a feminist geopolitics. *The Canadian Geographer* 54(2), 210–222.

Hyndman, J. (2003) Beyond either/or: a feminist analysis of September 11th. *ACME: An International E-Journal for Critical Geographies* 2(1), 1–13.

Inwood, J. and Tyner, J.A. (2011a) Geography's pro-peace agenda: an unfinished project. *ACME: An International E-Journal for Critical Geographies* 10(3), 442–57.

Inwood, J and Tyner, J.A. (2011b) *Non-Killing Geographies: Violence, Space and the Search for a More Humane Geography*. Center For Global Non-Killing, Honolulu, HA.

Iqtidar, H. and Lehman, D. (2012) Secularism and citizenship beyond the north Atlantic world. *Citizenship Studies* 16(8), 953–959.

Isin, E.F. (2002) *Being Political: Genealogies of Citizenship*. University of Minnesota Press, Minneapolis.

Isin, E. (2008) *Recasting the Social in Citizenship*. University of Toronto Press, Toronto.

Isin, E., Brodie, J., Juteau, D. and Stasiulis, D. (2008) Recasting the social in citizenship. In: Isin, E. (ed.) *Recasting the Social in Citizenship*. University of Toronto Press, Toronto, pp. 3–19.

Isin, E.F. and Nielsen, G.M. (2008) *Acts of Citizenship*. Zedbooks, New York.

Jacobs, J. and Fincher, R. (eds.) (1998) *Cities of Difference*. The Guildford Press, New York, London.

Jaffrelot, C. (1996) *Hindu Nationalist Movement, 1925–1992: Social and Political Strategies*. C. Hurst, London.

Jaffrelot, C. (2003) *India's Silent Revolution: The Rise of the Lower Castes in North India*. C. Hurst, London.

Jaffrelot, C. (2007) *Hindu Nationalism: A Reader*. Princeton University Press, Princeton, NJ.

Janoski, T. (1998) *Citizenship and Civil Society: A Framework of Rights and Obligations in Liberal, Traditional and Social Democratic Regimes*. Cambridge University Press, Cambridge.

Jasani, R. (2008) Violence, reconstruction and Islamic reform: stories from the Muslim ghetto. *Modern Asian Studies* 42(2–3), 431–456.

Jayal, N.G. (2011) The transformation of citizenship in India in the 1990s and beyond. In: Ruparelia, S., S. Reddy, J. Harriss et al. (eds.) *Understanding India's New Political Economy: A Great Transformation?* Routledge, London, pp. 141–156.

Jayal, N.G. (2013) *Citizenship and its Discontents: An Indian History.* Harvard University Press, Cambridge, MA.

Jeffery, R. and Jeffery, P. (1994) The Bijnor Riots, October 1990: collapse of a mythical special relationship? *Economic and Political Weekly* 29(10), 551–558.

Jeffery, R., Jeffrey, C. and Lerche, J. (eds.) (2014) *Failed Development and Identity Politics: India through the Lens of Uttar Pradesh.* Sage Publications, London.

Jeffrey, A. (2007) The politics of "democratization": lessons from Bosnia and Iraq. *Review of International Political Economy* 14(3), 444–466.

Jeffrey, C. (2008) Kicking away the ladder: student politics and the making of an Indian middle class. *Environment and Planning D: Society and Space* 26(3), 517–536.

Jeffrey, C. (2010) *Timepass: Youth, Class and the Politics of Waiting in India.* Stanford University Press, Stanford, CA.

Jeffrey, C., Jeffery, P. and Jeffery, R. (2008) *Degrees without Freedom: Education, Masculinities and Unemployment in North India.* Stanford University Press, Stanford, CA.

Jeffrey, C., and Lerche, J. (2000) Stating the difference: state, discourse and class reproduction in Uttar Pradesh, India. *Development and Change* 31(4), 857–878.

Jeffrey, R. (2000) *India's Newspaper Revolution: Capitalism, Politics and the Indian-Language Press, 1977–99.* C. Hurst, London.

Jeffrey, R. and Doron, A. (2012) Mobile-izing: democracy, organization and India's first "Mass Mobile Phone" elections. *The Journal of Asian Studies* 71, 63–80.

Johansen, J. (2006) Peace research needs to reorient. In: A Hunter (ed.) *Peace Studies in the Chinese Century.* Ashgate, Aldershot, pp. 31–38.

Johnson, D. (2012) Sri Lanka – a divided church in a divided polity: the brokerage of a struggling institution. *Contemporary South Asia.* 20(1), 77–90.

Jones, M. (2008) Recovering a sense of political economy. *Political Geography* 27(4), 377–399.

Jones, R. (2009) Geopolitical boundary narratives, the global war on terror and border fencing in India. *Transactions of the Institute of British Geographers NS* 34, 290–304.

Jones, R.D. (2010) Islam and the rural landscape: discourses of absence in west Wales. *Social and Cultural Geography* 11(8), 751–768.

Jutila, M, P. Samu and V. Tarja (2008) Resuscitating a discipline: an agenda for critical peace research Millennium. *Journal of International Studies* 36(3), 623–640.

Kabeer, N. (2007) *Inclusive citizenship: Meanings and Expressions.* Zed books, London.

Kakar, S. (1996) *The Colors of Violence: Cultural Identities, Religion and Conflict.* Chicago: Chicago University Press.

Kaldor, M. (1999) *New and Old Wars: Organized Violence in a Global Era.* Polity, Cambridge.

Kant, Immanuel (1795/1991) Zumewigen frieden [Perpetual peace]. In: Hans Reiss (ed.) *Kant's Political Writings*, 2nd edition. Cambridge University Press, Cambridge, pp. 93–130.

Kapila, S. (2010) A history of violence. *Modern Intellectual History* 7(2), 437–457.

Katz, C. (1994) Playing the field: Questions of fieldwork in geography. *The Professional Geographer* 46(1), pp. 67–72.

Katz, C. (2004) *Growing Up Global: Economic Restructuring and Children's Everyday Lives.* University of Minnesota Press, Minneapolis.

Kearns, G. (2009) *Geopolitics and empire: The Legacy of Halford Mackinder.* Oxford University Press, Oxford.

Khalidi, O. (2006) *Muslims in Indian Economy.* Three Essays, Gurgaon.

Khan, R. and S. Mittal (1984) The Hindu–Muslim Riot in Varanasi and the Role of the Police. In: Engineer, A.A. (ed.) *Communal Riots in Post Independence India.* Orient Longman, Bombay, pp. 305–312.

Khan, Y. (2007) *The Great Partition: The Making of India and Pakistan.* Yale University Press, New Haven, CT.

Kingdon, G.G. and M. Muzammil (2001a) A political economy of education in India – I. The Case of UP. *Economic and Political Weekly* 36(32), 3052–3063.

Kingdon, G.G. and M. Muzammil (2001b) The political economy of education in India – II. The case of UP. *Economic and Political Weekly* 36(33), 3178–3185.

Kliot, N. and S. Waterman (eds.). (1991) *Political Geography of Conflict and Peace.* Belhaven Press, London.

Knorringa, P. (1999) Artisan labour in the Agra footwear industry: Continued informality and changing threats. In: Parry, J.P., J. Breman and K. Kapadia (eds.) *The Worlds Of Indian Industrial Labour.* Sage Publications, New Delhi, pp. 303–328.

Knutsen, T.L. (1997) *A History of International Relations Theory: An Introduction.* Manchester University Press, Manchester.

Kobayashi, A. (2009). Geographies of peace and armed conflict: Introduction. *Annals of the Association of American Geographers* 99(5), 819–826.

Kohli, A. (1990) *Democracy and Discontent: India's Growing Crisis of Governability.* Cambridge University Press, Cambridge.

Koopman, S. (2008) Imperialism Within: Can the master's tools bring down empire? *ACME: An International E-Journal for Critical Geographies* 7(2), 283–307.

Koopman, S. (2011a) Alter-geopolitics: Other securities are happening. *Geoforum* 42(3), 274–284.

Koopman, S. (2011b) Let's take peace to pieces. *Political Geography* 30(4), 193–194.

Koopman, Sara (2014) Making space for peace: international protective accompaniment in Columbia. In: McConnell, F., N. Megoran and P. Williams (eds.) *Geographies of Peace.* I.B. Tauris, London, pp. 109–130.

Kumar, N. (1988) *The Artisans of Banaras: Popular Culture, Power and Identity, 1880–1986.* Orient Longman, New Delhi.

Kumar, N. (2007) *The Politics of Gender, Community, and Modernity: Essays On Education in India.* Oxford University Press, New Delhi.

Kumar Singh, A. 2007. The Economy of Uttar Pradesh since the 1990s: Economic stagnation and fiscal crisis. In: S Pai (ed.) *Political Process in Uttar Pradesh: Identity and Economic Reforms and Governance.* Pearson Longman, Delhi, pp. 273–294.

Kumar, V. (2007) Behind the BSP victory. *Economic and Political Weekly* 42(24), 2237–2239.

Kurlansky, M. (2006) *Nonviolence: The History of a Dangerous Idea*. Random House, New York.

Laliberte, N. (2014) Building peaceful geographies in and through systems of violence. In: McConnell, F., N. Megoran and P. Williams (eds.) *Geographies of Peace*. I.B. Tauris, London, pp. 47–65.

Lall, M. (2005) Indian education policy under the NDA government. In: Adeney, K. and L. Saez (eds.) *Coalition Politics and Hindu nationalism*. Routledge, London, pp.153–170.

Lau, T. (2009) Tibetan fears and Indian foes: Fears of cultural extinction and antagonism as discursive strategy. *Explorations in Anthropology* 9(1), 81–90.

Lazar, S. (2008) *El Alto, Rebel City: Self and Citizenship in Andean Bolivia*. Duke University Press, Durham, NC.

Lederach, J.P. (1997) *Building Peace: Sustainable Reconciliation in Divided Societies*. United States Institute of Peace Press, Washington DC.

Lefebvre, H. (1991) *The Production of Space*. Blackwell, Oxford.

Lefebvre, H. (1996) *Writings on Cities: Henri Lefebvre*. Blackwell, Oxford.

Lerche, J. (1999) Politics of the poor: Agricultural labourers and political transformations in Uttar Pradesh. In: Byres, T.J., K. Kapadia and J. Lerche (eds.) *Rural Labour Relations in India*. Taylor and Francis, London, pp. 182–241.

Levey, G.B. (2009) Secularism and religion in a multicultural age. In: Levey, G.B. and T. Modood (eds.) *Secularism, Religion and Multicultural Citizenship*. Cambridge University Press, Cambridge, pp. 1–24.

Levinas, E. (1978) *Otherwise than Being or Beyond Essence* (trans. A. Lingis). Kluwer, Dordrecht.

Lister, R. (1997) *Citizenship: Feminist perspectives*. Macmillan, Basingstoke.

Lobo, L. and B. Das (2006) *Communal Violence and Minorities*. Rawat Publications, New Delhi.

Low, S.M. and D. Lawrence-Zuniga (2003) *The Anthropology of Space and Place: Locating Culture*. Blackwell, Oxford.

Loyd, J. (2012) Geographies of peace and antiviolence. *Compass* 6(8), 477–489.

Lyon, D. (2004) Technology vs. Terrorism: Circuits of city surveillance since September 11, 2001. In: Graham, S. (ed.) *Cities, War, and Terrorism: Towards an Urban Geopolitics*. Blackwell, Oxford, pp. 297–311.

Madan, T.N. (1987) Secularism in its place. *The Journal of Asian Studies* 46(4), 747–759.

Mahmood, S. (2005) *Politics of Piety: The Islamic Revival and the Feminist Subject*. Princeton University Press, Princeton, NJ.

Malik, D. (1996) Three Riots in Varanasi: 1989–1990, 1991 and 1992. In: Bidwai, P., H. Mukhia and A. Vanaik (eds.) *Religion, Religiosity and Communalism*. Manohar, New Delhi, pp. 157–166.

Mamadouh, V. (2005) Geography and war, geographers and peace. In: Flint, C. (ed.) *The Geography of War and Peace: From Death Camps to Diplomats*. Oxford University Press, Oxford, pp. 26–60.

Mamdani, M. (1996) *Citizen and Subject: Contemporary Africa and the Legacy of Late Colonialism*. James Currey, London.

Mann, E.A. (1992) *Boundaries and Identities: Muslims, Work and Status in Aligarh.* Sage Publications, New Delhi.

Mannathukkaren, N. (2010) The "Poverty" of political society: Partha Chatterjee and the People's Plan Campaign in Kerala, India. *Third World Quarterly* 31(2), 295–314.

Manor, J. (1997) Parties and the Party System. In: Partha Chatterjee (ed.) *State and Politics in India.* Oxford University Press, Delhi, pp. 92–124.

Marshall, T.H. (1950) *Citizenship and Social Class.* Cambridge University Press, Cambridge.

Massey, D. (2005) *For Space.* Sage Publications, London.

Mathur, S. (2010) *Everyday Life of Hindu Nationalism. An Ethnographic Report.* Three Essays Publications, Delhi.

May, T. (2001) *Social Research: Issues, Methods and Process.* Open University Press, Buckingham.

Mayaram, S. (2005) Living together: Ajmer as a paradigm for the (South) Asian City. In: Hasan, M. and A. Roy (eds.) *Living Together Separately.* (Cultural India in History and Politics) Oxford University Press, Oxford, pp. 145–171.

McConnell, F. (2013) Citizens and refugees: constructing and negotiating Tibetan identities in exile. *Annals of the Association of American Geographers* 103(4), 967–983

McConnell, F. (2014) Contextualising and politicising peace: geographies of Tibetan satyagraha. In: McConnell, F., N. Megoran and P. Williams (eds.) *Geographies of Peace.* I.B. Tauris, London, pp. 131–150.

McConnell, F, N. Megoran and P. Williams (eds.) (2014). *Geographies of Peace.* I.B. Tauris, London.

McNay, L. (2000) *Gender and Agency: Reconfiguring the Subject in Feminist and Social Theory.* Polity Press, Cambridge.

McNay, L. (2004) Agency and experience: gender as a lived relation. *Sociological Review* 52(2), 173–190.

Medhasananda, S. (2002) *Varanasi at the Crossroads.* The Ramakrishna Mission Institute of Culture, Kolkata.

Megoran, N. (2005) The critical geopolitics of danger in Uzbekistan and Kyrgyzstan. *Environment and Planning D: Society and Space* 23(4), 555–580.

Megoran, N. (2006) For ethnography in political geography: Experiencing and re-imagining Ferghana Valley boundary closures. *Political Geography* 25(6), 622–640.

Megoran, N. (2010) Towards a geography of peace: Pacific geopolitics and evangelical Christian Crusade apologies. *Transactions of the Institute of British Geographers* NS 35(3), 382–398.

Megoran, N. (2011) War and peace? An agenda for peace research and practice in geography. *Political Geography* 30(4), 178–189.

Megoran, N. (2014) Migration and peace: the transnational activities of Bukharan Jews. In: McConnell, F., N. Megoran and P. Williams (eds.) *Geographies of Peace.* I.B. Tauris, London, pp. 212–228.

Mehta, U.S. (2010) The social question and the absolutism of politics. *Seminar* 615.

Menon, N. (2007) Living with Secularism. In: Needham, A.D. and R.S. Rajan (eds.) *Crisis of Secularism in India.* Duke University Press, Durham, NC, pp. 118–140.

Menon, N. (2011) The Ayodhya judgment: What next? *Economic and Political Weekly* 46(31), 81–89.

Michelutti, L. (2008) *The Vernacularisation of Democracy. Politics, Caste and Religion in India*. Routledge, London and Delhi.

Mills, S. (2009) Citizenship and Faith: Muslim scout groups. In: Phillips, R. (ed.) *Muslim Spaces of Hope: Geographies of Possibility in Britain and the West*. Zed Books, London, pp. 85–103.

Milton, G. (2009) *Paradise Lost: Smyrna 1922 The Destruction of Islam's City of Tolerance*. Hodder and Stoughton, London.

Mines, M. (1972) *Muslim Merchants: The Economic Behaviour of an Indian Muslim Community*. Sri Ram Centre for Industrial Relations and Human Resources, New Delhi.

Miraftab, F. and S. Wills (2005) Insurgency and spaces of active citizenship: The story of Western Cape anti-eviction campaign in South Africa. *Journal of Planning Education and Research* 25(2), 200–217.

Mitchell, D. (2003) *The Right to the City: Social Justice and the Fight for Public Space*. Guildford Press, New York.

Mitchell, K. (2011) Marseille's not for burning: comparative networks of integration and exclusion in two French cities. *Annals of the Association of American Geographers* 101(2), 404–423.

Mitchell, T. (1999) The limits of the state: beyond statist approaches and their critics. *The American Political Science Review* 85(1), 77–96.

Modood, T. (1994) Establishment, multiculturalism and British citizenship. *Political Quarterly* 65(1), 53–73.

Modood, T., A. Triandafyllidou and R. Zapata-Barrero (eds.) (2006) *Multiculturalism, Muslims and Citizenship*. Routledge, London.

Mohsini, M. (2010) Crafts, artisans and the nation-state in India. In: Clark-Deces, I. (ed.) *A Companion to the Anthropology of India*. Wiley-Blackwell, Oxford, pp. 186–201.

Moran, J. (2008) *Policing the Peace in Northern Ireland: Politics, Crime and Security after the Belfast Agreement*. Manchester University Press, Manchester.

Morris Jones, W.H. (1966) Dominance and dissent: Their inter-relations in the Indian party system. *Government and Opposition* 1(4), 451–466.

Morris Jones, W.H. (1978) *Politics mainly Indian*. Orient Longman, Madras.

Mukerjee, S. (2004) Hard times for sari weavers. *BBC News*. November 8, 2004.

Mukerji, D. (2006) Popularity test ride: Advani's yatra has few supporters in the BJP. *The Week*. April 16, 2006.

Nagle, J. and M.A.C. Clancy (2010) *Shared Society or Benign Apartheid? Understanding Peace-Building in divided Societies*. Palgrave Macmillan, Basingstoke.

Nanda, M. (2009) *The God Market: How Globalization is Making India more Hindu*. Random House, Delhi.

Nandy, A. (1998) The politics of secularism and the recovery of religious tolerance. In: Bhargava, R. *Secularism and its Critics*. Oxford University Press, Oxford, pp. 321–344.

Nandy, A. with S. Trivedy, S. Mayaram and A. Yagnik (1995) *Creating a Nationality: the Ramjanmabhumi Movement and Fear of the Self*. Oxford University Press, Delhi.

Nardin, T. (1996) *The Ethics of War and Peace: Religious and Secular Perspectives*. Princeton University Press, Princeton, NJ.

Navaro-Yashin,Y. (2003) "Life is dead here": Sensing the political in "no man's land". *Anthropological Theory* 3(1), 107–125.

Naylor, S. and J. Ryan (2002) The mosque in the suburbs: negotiating religion and ethnicity in South London. *Social and Cultural Geography* 3(1), 39–60.

Nielsen, G.M. (2008) Answerability with cosmopolitan intent: An ethics-based politics for acts of citizenship. In: Engin F. Isin and Greg M. Nielsen (eds.) *Acts of Citizenship*. Zedbooks, New York, pp. 266–286.

Nyers, P. (2007) Why citizenship studies? *Citizenship Studies* 11(1), 1–4.

Ó Tuathail, G. (2005) Being geopolitical: comments on Engin Isin's Being Political: Genealogies of citizenship. *Political Geography* 24(3), 365–372.

Olsen, K.D. (2003) "We all eat pickles, don't we?" Negotiating identity in the city of Ajmer. Doctoral Dissertation, Department of Anthropology. Syracuse, IL: Syracuse University.

Olsen, K.D. (2005) Disrupting an almost seamless discourse: Working-class Muslim women's accounts of a communal clash and curfew in the city of Ajmer. In: Hasan, Z. and R. Menon (eds.) *In a Minority: Essays on Muslim women in India*. Oxford University Press, Oxford, pp. 333–369.

Ong, A. (1999) *Flexible Citizenship: The Flexible Logics of Transnationality*. Duke University Press, Durham, NC.

Oza, R. (2006) The geography of right wing violence in India. In: Gregory, D. and A. Pred (eds.) *Violent Geographies: Fear, Terror and Political Violence*. Routledge, London.

Pai, S. (2007a) Introduction. In: Pai, S. (ed.) *Political Process in Uttar Pradesh: Identity, Economic Reforms and Governance*. Pearson Longman, Delhi, pp. xxv–xlviii.

Pai, S. (2007b) *Political process in Uttar Pradesh: Identity, Economic Reform and Governance*. Pearson Longman, New Delhi.

Pain, R. (2009) Globalised fear? Towards an emotional geopolitics. *Progress in Human Geography* 33(4), 1–21.

Pain, R. and S.J. Smith (eds.) (2008) *Fear: Critical Geopolitics and Everyday Life*. Aldershot, Ashgate.

Painter, J. (2006) Prosaic geographies of stateness. *Political Geography* 25(7), 752–774.

Pallister-Wilkins, P. (2011) The Separation Wall: A symbol of power and a site of resistance? *Antipode* 43(4), 1851–1882.

Pandey, G. (1983) The bigoted Julaha. *Economic and Political Weekly* 18(5), 19–28.

Pandey, G. (1990) *The construction of communalism in colonial North India*. Oxford University Press, Delhi.

Pandey, G. (1999) Can a Muslim be an Indian? *Comparative Studies in Society and History* 41(4), 608–629.

Pandey, G. (2001) *Remembering Partition: Violence, Nationalism and History in India*. Cambridge University Press, Cambridge.

Pandey, G. (2005) *Routine Violence: Nations, Fragments, Histories*. Stanford University Press, Stanford, CA.

Pangle, T.L. and P.J. Ahrensdorf (1999) *Justice among Nations: On the Moral Basis of Power and Peace*. University Press of Kansas, Lawrence.

Papanek, H. (1964) The woman field worker in a purdah society. *Hutman Organisation* 23(2), 160–163.

Parry, J. (1994) *Death in Banaras*. Cambridge University Press, Cambridge.

Patel, G. (2010) Idols in law. *Economic and Political Weekly* 45(50), 47–52.

Peach, C., V. Robinson, V. and S.J. Smith (eds.) (1981) *Ethnic Segregation in Cities*. Croom Helm, London.

Peake, L. and A. Kobayashi (2002) Policies and practices for anti-racist geography at the millennium. *The Professional Geographer* 54(1), 50–61.

People's Union for Civil Liberties (PUCL) (2000) Police attacks on Muslims during Muharram. Uttar Pradesh chapter of the People's Union for Civil Liberties, Varanasi.

People's Union for Civil Liberties (PUCL) (2008) Shahbaz Ahmed wrongly charged in connection with serial bomb blasts in Jaipur. People's Union for Civil Liberties (PUCL), Varanasi and People's Union for Human Rights (PUHR), New Delhi.

Pepper, D. and A. Jenkins (1983) A call to arms: geography and peace studies. *Area* 15(3), 202–208.

Phillips, R. (2009) Bridging east and west: Muslim-identified activists and organisations in the UK anti-war movements. *Transactions of the Institute of British Geographers* NS 34(4), 506–520.

Phillips, R. (2010) Common ground? Anti-imperialism in the UK anti-war movements. In: Dodds, K. and A. Ingram (eds.) *Spaces of In/Security: New Geographies of the War on Terror*. Ashgate, Aldershot, pp. 239–257.

Phillips, R. and J. Iqbal (2009) Muslims and the anti-war movements. In: Phillips, R. (ed.) *Muslim Spaces of Hope: Geographies of Possibility in Britain and the West*. Zed Books, London, pp. 163–178.

Pradhan, S. (2006) Ancient Varanasi keeps its peace, proves its mettle. *Combating Communalism* 12(114).

Prakash, G. (2002) The urban turn. *Sarai Reader* 2(7) (The Cities of Everyday Life).

Prime Minister's High Level Committee (2006) *Social, Economic and Educational Status of the Muslim Community of India*. (Sachar Report) Cabinet Secretariat, Government of India, New Delhi.

Pugh, M. (2004) Peacekeeping and critical theory. *International Peacekeeping* 11(1), 39–58.

Pugh, M. (2005) The political economy of peacebuilding: a critical theory perspective. *International Journal of Peace Studies* 10(2), 23–42.

Pugh, M, N. Cooper and M. Turner (2011) *Whose Peace? Critical Perspectives on the Political Economy of Peacebuilding*. (New Security Challenges). Palgrave Macmillan, Basingstoke.

Putnam, R. (2000) *Bowling Alone: The Collapse and Revival of American Community*. Simon and Schuster, New York.

Raghuraman, S. (2004) Better Majboor than Mazboot. *Times News Network*. New Delhi.

Rai, M.M. (2015) Ayodhya litigants look to end decades-old Ram Mandir-Babri Masjid dispute. *Economic Times* April 11, 2015.

Rajagopal, A. (2001) *Politics after Television: Hindu Nationalism and the Reshaping of the Public in India*. Cambridge University Press, Cambridge.

Rajalakshmi, T.K. (2006) Flare-up at midnight. *Frontline* (23)3.

Raman, V. (2010) The warp and the weft: community and gender identity among the weavers of Banaras. Routledge, New Delhi.

Ramesh, R., V. Dodd, J. Burke and P. Beaumont (2008) Mumbai terror attacks: India fury at Pakistan as bloody siege is crushed. *The Guardian*, November 30.

Rawat, V.B. (2003) *Press and Prejudice: An Insightful Analysis of Hindi Media*. Institute of Objective Studies, New Delhi.

Richmond, O. (2005) *The Transformation of Peace*. Palgrave Macmillan, Basingstoke.

Richmond, O. (2008) *Peace in International Relations*. Routledge, London.

Richmond, O. (2009) Becoming liberal, unbecoming liberalism: liberal–local hybridity via the everyday as a response to the paradoxes of liberal peacebuilding *Journal of Intervention and Statebuilding* 3(3), 324–344.

Richmond, O. (2010) Resistance and the post-liberal peace. *Millennium – Journal of International Studies* May 2010 38(3), 665–692.

Rigg, J. (2007). *An Everyday Geography of the Global South*. Routledge, London.

Ring, L. (2006) *Zenana: Everyday Peace in a Karachi Apartment Building*. Indiana University Press, Bloomington.

Roberts, R. (1999) Another member of our family: aspects of television culture and social change in Varanasi, North India. Doctoral Dissertation, Department of Anthropology University of Edinburgh.

Roberts, S. (2003) Creative television in the Siti of Varanasi. In: Jeffrey, R. and J. Lerche (eds.) *Social and Political Change in Uttar Pradesh: European Perspectives*. Manohar, New Delhi.

Robinson, J. (2002) Global cities and world cities: a view from off the map. *International Journal of Urban and Regional Research* 26(3), 531–554.

Roy, T. (1993) *Artisans and Industrialisation: Indian Weaving in the Twentieth Century*. Oxford University Press, New Delhi.

Roy, B. (1994) *Some Trouble with a Cow: Making Sense of Social Conflict*. University of California Press, Berkeley.

Roy, A. (2002) *City Requiem, Calcutta: Gender and the Politics of Poverty*. University of Minnesota Press, Minneapolis.

Roy, A. (2004) *An Ordinary Person's Guide to Empire*. South End Press, Boston.

Roy, K. (2010) Issues of faith. *Economic and Political Weekly* 45(50), 53–60.

Roy, A. (2011) *Mapping Citizenship in India*. Oxford University Press, New York.

Ruthven, O. (2008) Metal and morals in Moradabad: Perspectives on ethics in the workplace across a global supply chain. Doctoral Dissertation, International Development Centre. University of Oxford, Oxford.

Saltzman, D. (2006) *Shooting Water*. Penguin, New Delhi.

Sandercock, L. (1998) *Towards Cosmopolis: Planning for Multicultural Cities* John Wiley, Chichester.

Scanlon, T.M. (1998) The difficulty of tolerance. In: Bhargava, R. (ed.) *Secularism and its Critics*. Oxford University Press, Oxford, pp. 54–72.

Schenk-Sandbergen, L. (1998) Gender in field research: Experiences in India. In: M.E. Thapan (ed.) *Anthropological Journeys: Reflections on Fieldwork.* Orient Longman, Hyderabad, pp. 267-292.

Scheper-Hughes, N. (1993) *Death without Weeping. The Violence of Everyday Life in Brazil.* University of California Press, Berkeley.

Scheper-Hughes, N. and P. Bourgois (2004) *Violence in War and Peace: An Anthology.* Wiley-Blackwell, Oxford.

Schutte, S. (2006) The social landscape of the washermen in Banaras. In: Gaenszle, M. and J. Gengnagel (eds.) *Visualising Space in Banaras: Images, Maps, And The Practice Of Representation.* Harrasssowitz Verlag, Wiesbaden, pp. 279–302.

Scott, J. (1985) *Weapons of the Weak: Everyday Forms of Peasant Resistance.* Yale University Press, New Haven, CT.

Scott, J. (1990) *Domination and the Arts of Resistance: Hidden Transcripts.* Yale University Press, New Haven, CT.

Scott, J. (2009) *The Art of not being Governed: An Anarchist History of Upland Southeast Asia.* Yale University Press, New Haven, CT.

Seabrook, J. and I.A. Siddiqui (2011) *People without history: India's Muslim ghettos.* Pluto Press, London.

Searle-Chatterjee, M. (1981) *Reversible Sex Roles: The Special Case of Benares Sweepers.* Pergamon, Oxford.

Searle-Chatterjee, M. (1994a). Caste, religion and other identities. In: Searle-Chatterjee, M. and U. Sharma (eds.) *Contextualising Caste.* Blackwell Publishers, Oxford, pp. 147–169.

Searle-Chatterjee, M. (1994b). Wahabi sectarianism among Muslims of Banaras. *Contemporary South Asia* 3(2), 83–93.

Secor, A. (2004) "There is an Istanbul that belongs to me": Citizenship, space and identity in Turkey. *Annals of the Association of American Geographers* 94(2), 352–368.

Selby, J. (2011) The political economy of peace processes. In: Pugh, M.N. Cooper and M. Turner. *Whose Peace? Critical Perspectives on the Political Economy of Peacebuilding* (New Security Challenges). Palgrave Macmillan, Basingstoke.

Seligman, A. (1992) *The Idea of Civil Society.* Princeton University Press, Princeton, NJ.

Sen, A. (1998) Secularism and its discontents. In: R. Bhargava (ed.) *Secularism and its Critics.* Oxford University Press, New Delhi, pp. 454–485.

Sen, A. (2007) *Identity and Violence: The Illusion of Destiny.* Penguin, London.

Sen, A. (2009) *The Idea of Justice.* Allen Lane, London.

Sennett, R. (1994) *Flesh and Stone: The Body and the City in Western Civilisation.* Faber and Faber, London.

Sennett, R. (2011) *Together: The Rituals, Pleasures and Politics of Cooperation.* Yale University Press, New Haven, CT.

Shah, A. (2007) 'Keeping the state away': democracy, politics, and the state in India's Jarkhand. *Journal of the Royal Anthropological Institute* 13(1), 129–145.

Shah, A. (2010) *In the Shadows of the State: Indigenous Politics, Environmentalism and Insurgency in Jharkhand, India.* Duke University Press, Durham, NC.

Shani, O. (2010) Conceptions of citizenship in India and the 'Muslim Question'. *Modern Asian Studies* 44(1), 145–173.

Shapiro, I. and S. Bedi (2007) *Political Contingency: Studying the Unexpected, the Accidental and the Unforeseen.* New York University Press.

Sharp, J. (2011) Subaltern geopolitics: Introduction. *Geoforum* 42(3), 271–273.

Showeb, M. (1994) *Silk Handloom Industry of Varanasi: A Study of Socio-Economic Problems of Weavers.* Ganga Kaveri Publishing House, Varanasi.

Sikand, Yoginder (2005) On the boil: The recent eruption in Mau is symptomatic of the fact that eastern UP is sitting on the mouth of a communal volcano. *Combating Communalism* 12(112).

Singh, N. and Srinivasan, T.N. (2005) Indian federalism, economic reform and glob-alization. In: J. Wallack and T.N. Srinivasan (eds.) *Federalism, Economic Reform and Globalization.* Cambridge University Press, Cambridge, pp. 301–363.

Singh, R.P.B. and P.S. Rana (2002) *Banaras Region: A Spiritual and Cultural Guide.* Indica, Varanasi.

Sinha, B.K. (2005) *Development Commissioner (Handlooms).* Ministry of Textiles. Government of India Office of the Development Commissioner for Handlooms, New Delhi.

Smith, S. (1989) Society, space and citizenship: a human geography for the "New Times"? *Transactions of the Institute of British Geographers NS* 14(2), 144–156.

Sparke, M. (2008) Political geography: political geographies of globalisation: III: resistance. *Progress in Human Geography* 32(3), 423–440.

Sponsel, L.E. and T. Gregory (1994) *The Anthropology of Peace and Nonviolence.* Lynne Reinner Publishers, Boulder, CO.

Springer, Simon (2011) Violence sits in places? Cultural practice, neoliberal ratio-nalism, and virulent imaginative geographies. *Political Geography* 30(2), 90–98.

Srivastava, R. (2009) *The Challenge of Employment in India. An Informal Economy Perspective,* Volume 1. The National Commission for Enterprises in the Unorganized Sector. New Delhi.

Stadler, N., E. Lomsky-Feder and E. Ben-Ari (2008) Fundamentalism's encounter with citizenships: The Haredim in Israel. *Citizenship Studies* 12(3), 215–231.

Staeheli, L.A. (2005) Machines without operators and genealogies without people: Comments on Engin Isin's Being Political. *Political Geography* 24(3), 349–353.

Staeheli, L.A. (2008) Citizenship and the problem of community. *Political Geography* 27(1), 5–21.

Staeheli, L.A. and E. Kofman (2004) Mapping gender, making politics: Feminist perspectives on political geography. In: Staeheli, L.A., E. Kofman L. Peake (eds.) *Mapping Women, Making Politics: Towards Feminist Political Geographies.* Routledge, London, pp. 1–14.

Ståhlberg, P. (2002) *Lucknow Daily: How a Hindu Daily Constructs Society.* Stockholm Studies in Anthropology, Stockholm.

Stepputat, F. (2004) Violence, sovereignty and citizenship in postcolonial Peru. In: Hansen, T.B. and F. Stepputat (eds.) *Sovereign Bodies: Citizens, Migrants and States in the Postcolonial World.* Princeton University Press, Princeton, NJ, pp. 61–81.

Stolberg, I. (1965) Geography and peace research. *The Professional Geographer* 17(4), 9–12.

Subramanian, A. (2003) Modernity from below: local citizenship on the South Indian coast. *International Social Science Journal*, 175 (March), 135–144.

Tambiah, S. (1996) *Levelling Crowds: Ethnonationalist Conflicts and Collective Violence in South Asia*. University of California Press, Berkeley.

Tarlo, E. (1996) *Clothing Matters: Dress and Identity in India*. C. Hurst, London.

Tejani, S. (2008) *Indian Secularism: A Social and Intellectual History 1890–1950*. Indiana University Press, Bloomington.

Thakur, R. (2011) *The Responsibility to Protect: Norms, Laws, and the Use of Force in International Politics*. Routledge, London.

Thomas, D.A. and R.J. Ely (2001) Cultural diversity at work: the effects of diversity perspectives on work group processes and outcomes. *Administrative Science Quarterly* 46(2), 229–273.

Thompson, K.W. (2006) Peace Studies: Social movement or intellectual discipline? *Ethics and International Affairs* 4(1), 163–174.

Tonkiss, F. (2003) The ethics of indifference: community and solitude in the city. *International Journal of Cultural Studies* 6(3), 297–311.

Tse, J.K.H. (2014) Grounded theologies: "Religion" and the "secular" in human geography. *Progress in Human Geography* 38(2), 201–220

Tully, M. (2007) *India's Unending Journey: Finding a Balance in a Time of Change*. Random House, London.

Tyner, J. and J. Inwood (2014) Violence as fetish: Geography, Marxism, and dialectics. *Progress in Human Geography*. 38(6), 771–784.

Vaghela, S. (2004) Speech of Shri. Shankar Vaghela, Minister of Textiles Union Minister of Textiles. New Delhi: All India Handloom Fabrics Marketing Co-op Society Limited.

Valentine, Gill (2008) Living with difference: Reflections on geographies of encounter. *Progress in Human Geography* 32(3), 323–337.

van der Veer, Peter (1994) *Religious Nationalism: Hindus and Muslims in India*. California University Press, Berkeley.

Varadarajan, S. (2002) *Gujarat: The Making of a Tragedy*. Penguin Books India, New Delhi.

Varma, P.K. (1998) *The Great Indian Middle Class*. Penguin Books, New Delhi.

Varma, P.K. (2005) *Being Indian: Inside the Real India*. William Heinemann, London.

Varshney, A. (2001) Ethnic conflicts and civil society: India and beyond. *World Politics* 53(3), 362–398.

Varshney, A. (2002) *Ethnic Conflict and Civic Life: Hindus and Muslims in India*. Oxford University Press, New Delhi.

Venkatesan, S. (2009) *Craft Matters: Artisans, Development and the Indian Nation*. Orient BlackSwan, Hyderabad.

Walzer, M. (1997) *On Toleration*. Yale University Press, New Haven, CT.

Werbner, P. and N. Yuval Davis (1999) Women and the new discourse of citizenship. In: Werbner, P. and N. Yuval Davis (eds.) *Women, Citizenship and Difference*. Zed Books, London, pp. 1–38.

Wilkinson, S.I. (2004) *Votes and Violence: Electoral Competition and Communal Riots in India*. Cambridge University Press, Cambridge.

Wilkinson-Weber, Clare M. (1999) *Embroidering Lives: Women's Work and Skill in the Lucknow Embroidery Industry*. State University of New York Press, Albany.

Williams, P. and F. McConnell (2011) Critical geographies of peace. *Antipode* 43(4), 927–931.

Williams, P., B. Vira and D. Chopra (2011) Marginality, agency and power: Experiencing the state in contemporary India. *Pacific Affairs* 84(1), 7–23.

Wirsing, R.G. (1998) *India, Pakistan and the Kashmir Dispute: On Regional Conflict and its Resolution.* Macmillan, Basingstoke.

Wolf, D.L. (1996) Situating feminist dilemmas in fieldwork. In: Wolf, D.L. and C.D. Diana (eds.) *Feminist Dilemmas in Fieldwork.* Westview, Boulder, CO, pp. 1–55.

Woon, C.Y. (2011) Undoing violence, unbounding precarity: Beyond the frames of terror in the Philippines. *Geoforum* 42(3), 285–296.

Woon, C.Y. (2014) Precarious geopolitics and the possibilities of nonviolence. *Progress in Human Geography* 38(5), 654–670.

Woon, C.Y. (2015) "Peopling" geographies of peace: the role of the military in peace-building in the Philippines. *Transactions of the Institute of British Geographers.* 40(1), 14–27.

World Bank, The (2002) *Poverty in India: The Challenge of Uttar Pradesh.* World Bank, Washington, DC.

World Bank, The (2006) *Monitoring Poverty in Uttar Pradesh: A Report on the Second Poverty and Social Monitoring Survey (PSM-II).* Directorate of Economics and Statistics Planning Department, Government of Uttar Pradesh and the World Bank, Joint Report, June 2006.

Yadav, Y. (2009) Rethinking social justice. *Seminar* 601(September)

Yiftachel, O. (2006) *Ethnocracy: Land, and the Politics Of Identity Israel/Palestine.* University of Pennsylvania Press, Philadelphia.

Yiftachel, O. and H. Yacobi (2002) Urban ethnocracy: ethnicization and the production of space in an Israeli "mixed city." *Environment and Planning D: Society and Space* 21(6), 673–693.

Young, I.M. (1990) *Justice and the Politics Of Difference.* Princeton University Press, Princeton, NJ.

Yuval-Davis, N. and P. Werbner (1999) *Women, Citizenship and Difference: Postcolonial Encounters.* Zed Books, London.

Newspapers

Dainik Jagran July 25, 2008
Economic Times April 11, 2015
Times of India September 4, 2005
Times of India March 8, 2006
Times of India March 10, 2006
Times of India March 11, 2006
Times of India November 16, 2008
Times of India February 11, 2009
The Hindu April 20, 2006
The Hindu June 21, 2009
The Hindu June 29, 2009
The Wall Street Journal November 25, 2007
Thaindian.com November 14, 2007

Index

Everyday Peace?: Politics, Citizenship and Muslim Lives in India, First Edition. Philippa Williams.
© 2015 John Wiley & Sons, Ltd. Published 2015 by John Wiley & Sons, Ltd.